PRAISE FOR
THE WOODCHIP HANDBOOK

'In a world desperate for solutions that will allow us to get off the agro-chemical treadmill while tackling the climate challenges ahead, woodchip steps forward. Working with Ben, I've learned much over the last few years, and am amazed by the power of this humble resource. This trailblazing book will be invaluable in firing both our imagination and understanding of woodchip's potential to grow and farm better.'

—HELEN BROWNING,
chief executive, Soil Association

'I did not know the world needed a whole book on woodchip, but from the first chapter I could not put this book down. Raskin's thorough, informed and well-referenced research is going to be a must for anyone who cares about soil. Woodchip has such potential to help us create more local, sustainable ways of growing, and this handbook is the resource needed to dispel myths and find the best method and scale for everyone from gardeners to large-scale farms.'

—ALYS FOWLER, author of
The Edible Garden and *The Thrifty Gardener*

'I use woodchip quite extensively in my garden and have for decades. It has been an important part of increasing our soil organic matter and carbon storage. *The Woodchip Handbook* addresses its many uses and all the technical questions of how to acquire and manage it. If you want to learn more about woodchip for your yard or farm, look no further.'

—ERIC TOENSMEIER, author of
The Carbon Farming Solution and *Perennial Vegetables*

'Ben Raskin's wide-ranging work in varied locations makes *The Wood-chip Handbook* a really useful overview of the possibilities afforded by woodchip. It's good for soil and good for plants, once you understand how it can work best for you, as he explains in this book.'

—CHARLES DOWDING, author of
Charles Dowding's No Dig Gardening

'In a time when the farming and gardening literature seems to be overflowing with information about soil mineralization with rock dusts, cover crop smorgasbords, foliar feeding, liquid carbon pathways and rotational grazing, it is great relief to finally have, in one place, a treasure that describes the missing piece of the complete soil creation process: coarse woody debris! You won't find a better, more complete summary of the how, the why and the overall importance of wood-chip in the process of building the ideal fertile soil for your garden, homestead or farm. Bravo, Ben Raskin!'

—MARK SHEPARD, author of *Restoration Agriculture*

'Ben Raskin's latest book, on the many, many uses and benefits of woodchip, is the definitive guide to the subject. It's something every climate-change-savvy gardener must read.'

—KIM STODDART, editor, *The Organic Way* magazine;
coauthor of *The Climate Change Garden*

'A real "grower's" guide: comprehensive, informed and, quite possibly, revolutionary. Raskin is a practitioner of soil, and woodchip is his medi-cine. Through astute observation and keen experimentation, he unravels the burgeoning possibilities of woodchip for soil health and regeneration. The secret to woodchip, Raskin says, is fungi: "Woodchip without fungi is like a sea without fish." If the future of growing is fungi, this book is the prototype manual for practical application of its unique power.'

—MATT REES-WARREN, garden designer and
author of *The Ecological Gardener*

'Ben Raskin covers all the bases when it comes to woodchip. This very useful and abundant by-product improves the soil and the environment. Ben presents a wealth of information in a logical, clear and engaging manner. I've been a longtime user of woodchip; after reading this book, I will begin using even more of it in even more ways.'

—LEE REICH, author of
The Pruning Book and *Weedless Gardening*

'Woodchip is widely available in most cities and has much potential in container growing – as a sustainable source of nutrients, as an ingredient to help sustain the structure of compost over many years and as a brilliant mulch for fruit trees and other perennials. At last, thanks to Ben Raskin, we have the first authoritative, well-researched book on the subject, packed full of useful information and advice. A superb resource for anyone interested in the magic of woodchip.'

—MARK RIDSDILL SMITH, author of
The Vertical Veg Guide to Container Gardening

'Before picking up this book, I considered woodchip simply a nice thing to have around as mulch, but the deep dive Ben Raskin offers really makes the case that this material is an essential and focal part of so many aspects when stewarding healthy plants, soil, livestock and fungi. Readers will find the research and case studies fascinating, and the extensive details helpful in answering some of the most commonly asked questions. In the end, if we are to get serious in the act of regenerative farming and ecosystem repair, it's clear that woodchip will play a critical role and it's time we get more serious about utilizing it.'

—STEVE GABRIEL, Wellspring Forest Farm;
author of *Silvopasture*

'Tree ecosystems create long-term soil fertility. We tap into lignin potential and fungal doings in our own gardens and orchards by utilizing woodchip in a multitude of ways. Ben Raskin delves into the practical nuance of ramial chipped wood and more to inspire humus building efforts everywhere. Regenerate your soil with the woodchip!'

—MICHAEL PHILLIPS, author of
The Holistic Orchard and *Mycorrhizal Planet*

'This is a timely piece of work given the dramatic rise, access and popularity of woodchip technologies. Ben Raskin has done the subject proud: well researched and with a pragmatic approach, there is something for growers and gardeners of all shapes and sizes here. Woodchip is able to offer truly sustainable solutions to some of the problems facing horticulture, in particular the replacement of peat-based products. With woodchip we can now grow our own fertile soil and substrates. This manual gives some great information towards that goal.'

—IAIN TOLHURST, Tolhurst Organics

'Agriculture has been reinventing itself regularly over the past ten thousand years. *The Woodchip Handbook* is a practical guide for the next step in that transformation in our backyard gardens and orchards as well as on our largest farms. It's also a great read. Let the trees lead the way!'

—JOHN BUNKER, orchardist and author
of *Apples and the Art of Detection*

THE

WOODCHIP
HANDBOOK

Also by Ben Raskin

Zero-Waste Gardening

Bees, Bugs and Butterflies:
A Family Guide to Our Garden Heroes and Helpers

Grow: A Family Guide to Growing Fruits and Vegetables

Compost: A Family Guide to Making Soil from Scraps

The Community Gardening Handbook

THE
WOODCHIP
HANDBOOK

A COMPLETE GUIDE FOR
Farmers, Gardeners *and* Landscapers

BEN RASKIN

Chelsea Green Publishing
White River Junction, Vermont
London, UK

Disclaimer: Fungal spores present in composting woodchip can cause health problems if inhaled.
Consider wearing a mask when handling woodchip particularly in an enclosed situation.

Project Manager: Alexander Bullett
Developmental Editor: Benjamin Watson
Copy Editor: Susan Pegg
Proofreader: Laura Jorstad
Indexer: Lisa Himes
Designer: Melissa Jacobson
Page Layout: Abrah Griggs

Printed in the United Kingdom.
First printing October 2021.
10 9 8 7 6 5 4 3   22 23 24 25

Library of Congress Cataloging-in-Publication Data
Names: Raskin, Ben, 1969- author.
Title: The woodchip handbook: a complete guide for farmers, gardeners and landscapers / Ben Raskin.
Description: White River Junction, Vermont: Chelsea Green Publishing, [2021] | Includes bibliographi-
   cal references and index.
Identifiers: LCCN 2021036792 (print) | LCCN 2021036793 (ebook) | ISBN 9781645020486 (paperback) |
   ISBN 9781645020493 (ebook)
Subjects: LCSH: Wood chips. | Wood waste.
Classification: LCC HD9769.W38 R37 2021 (print) | LCC HD9769.W38 (ebook) | DDC
   338.4/7674—dc23
LC record available at https://lccn.loc.gov/2021036792
LC ebook record available at https://lccn.loc.gov/2021036793

Chelsea Green Publishing
85 North Main Street, Suite 120
White River Junction, Vermont USA

Somerset House
London, UK

www.chelseagreen.com

CONTENTS

INTRODUCTION

After more than a quarter of a century working in horticulture, there are things that still give me a visceral thrill. One is to see new buds and shoots in spring; those tender cotyledons emerging in the seed trays, bulbs tentatively poking their noses from the soil to sniff the warmer air and fruit buds bursting with potential on bare branches. Another is the magic that happens in a compost heap or woodpile. Somehow this material turns from leaf, wood or fetid vegetable waste into a sweet-smelling, friable crumb with the power to transform soil and plants. Working as a youngster at the stunning William Robinson designed Gravetye Manor gardens in Sussex in England, we used copious amounts of mushroom compost to mulch the ornamental borders. In hindsight, we overdid it and it was often not properly matured, but shovelling steaming piles of compost on a cold winter morning was one of my favourite jobs.

As I became more experienced and observant, I began to understand the importance of this biological cycling on soil health. I also noticed that while compost was widely recognised as useful and vital to plant and soil health, woodchip seemed to be less appreciated, apart from as a mulch for paths and landscaping. This nagging feeling of untapped possibilities gradually crystallised into something more concrete over the years. In a classic case of the Baader-Meinhof phenomenon (where you come across something new and then suddenly notice it everywhere, having never previously been aware of it), once I was on the lookout for practical applications of woodchip, I couldn't seem to avoid it. One of the perks of my work with the UK charity the Soil Association is that I meet and learn from many growers, farmers, advisers and researchers with far more skill and experience than I have, so there was plenty of opportunity to tick off the pages in my woodchip spotters guide; mulches, hedge management and propagation compost all built the case. Then I discovered ramial woodchip (usually known as 'ramial chipped wood'), which is made from young, freshly cut branches, and was surprised that its capacity to boost soil health had not been embraced by more farmers and gardeners.

It turns out, of course, that I am quite late to the party and that there are plenty of people who have been working with woodchip for decades, but for some reason it has not become mainstream. Perhaps the reliance of the dominant farming systems on artificial fertilisers, or the persistent misconception of nitrogen lock-up risk from using woodchip are to blame – the latter of which is discussed in detail in chapter 5. The global awakening to the importance of soil health and climate change perhaps provides the cue for woodchip to step into the limelight. We've still lots to learn about how to best use woodchip, and especially the potential for using single-species chip for particular purposes, but the research I have done for this book has only confirmed my practical experiments and observations that, with a few basic precautions, woodchip can only enhance your garden or farm.

The underlying premise for this book is that adding woodchip to our soils will give them long-term health benefits, but that we have numerous opportunities along that journey to enjoy the added perks of the material by using it for mulching, propagation and so on. Once we appreciate the full value of woodchip we can begin to design farming and growing systems around it, thus providing a valuable role for more trees in the farmed landscape. Rather than seeing tree management as an unprofitable and time-consuming chore, we can approach it as a profitable harvest from our plots and farms, with all the joy that harvesting food or flowers might give us.

CHAPTER ONE

What Is Woodchip and Where Does It Come From?

Wood is an abundant and renewable material that humans have exploited probably for as long as we have existed. We make useful objects from it, such as tools, ships, furniture and paper. We craft beautiful objects like sculptures and musical instruments. We also set fire to wood for cooking and warmth, amongst other things. There are certain species that are best suited to each use: oak for ships, yew for longbows, and maple, spruce and willow for the different parts of a Stradivarius violin. Though flint axes and saws have been around for many thousands of years it is only with the discovery of bronze and the smelting of iron, and the subsequent invention of metal axes and saws that cutting, chopping and processing trees became a whole lot easier. Once you have cut and milled the tree for its primary purpose, however, there are bits left over. The small branches, the bark and the other offcuts are often less valuable. Sawdust and wood shavings have always had a use, however, even if just for absorbing booze and blood on the floor of sleazy bars. Over time, no doubt, people observed what happened when chipped wood, perhaps combined with manure or human waste, was left for any length of time, and this led to an understanding of how it could be used in horticultural systems.

It was not until 1884 when Peter Jensen invented the first woodchipper in Germany, to help the Maasbüll local authorities deal with the timber produced from their public parks, that the material was produced on a significant scale. Since then, not only have these machines become the standard way to deal with woody tree waste, but a whole industry of producing chip for burning for heat and energy has developed. 'Woodchip' is, therefore, a very broad term that can refer to any woody material that has been broken up into small pieces. Since much of this chip is used in biomass boilers, most classification methods for woodchip are for this purpose and of little use in the context of growing crops and soil health. Chip for burning is standardised by chip size, ash and moisture content. The first of these may be helpful when choosing which machinery to use to chip or shred, or if you are using machinery to spread your material and need to gauge what size of chip will prevent clogging. However, ash and moisture are less useful measures, so we need to find other ways to assess our woodchip for agricultural purposes.

The two main considerations are the source of your chip and the tree species from which it has come. The age and/or part of the tree can also be important in some applications; for instance, we will be looking at ramial chipped wood, which is woodchip made from branches less than 7 centimetres (2¾ inches) diameter, for adding directly to soil. We also want to know whether our chip is purely bark, or whether it includes the whole tree. If we produce our own chip, we can control much of this. When relying on external suppliers, though, understanding the properties of different materials and the risks associated with them, and how they have been produced will help us to source and manage our resource.

While some of the uses of woodchip are still quite novel, such as making a propagating hotbed or improving soil health, their use in landscaping and gardening as a mulch is widespread. As a result, there are already some classifications used in this sector. They are not universally adopted, but they can give us some reassurance of quality. For example, one UK mulch supplier has created their own quality standard. They guarantee that their material is: '100% British mixed composted wood chip fines … The composted wood fines are screened to 10mm and composted for a minimum of 10 weeks. The fines are turned at least three times to aid the composting process.'[1]

In the absence of an agreed standard for horticultural woodchip, we need to critically assess any material used. Products of different quality may need to be handled or processed differently. For instance, we might compost a batch for longer, or sieve out the chip into two sizes with a different use for each. Unless heavily contaminated, pretty much all woodchip has some use, and I hope this book will give you the tools you need to be able to make that assessment and harness the amazing powers of woodchip.

Overview of Uses for Woodchip

I aim to cover a range of woodchip uses in both commercial horticulture and home gardens. While there are some instances that are relevant to only one of these, in most cases the principles are equally applicable even where the exact methods may depend on your location and scale.

Mulching

Mulching is the most common and best-understood use for woodchip in horticulture. Though occasionally abused in municipal planting or with artificially coloured chip, we get a whole range of benefits from adding woodchip as a surface covering to soil. There are risks, of course, and making sure we use the right material in the right situation is important. Let's briefly run through some of woodchip mulch's superpowers, all of which will be covered in more detail later in the book.

WEED CONTROL

Covering soil with a material free of weed seed that cuts out light will almost entirely eliminate annual weeds from your soil surface. It can also weaken some perennial weeds or at least make them easier to remove. Since weeding is one of the most time-consuming jobs in horticulture, especially in wetter climates, we can see why mulches are an attractive option.

There are a whole range of materials that can be used in this way: cardboard, sterile compost and plastic mulches (woven or solid). An organic material used for this purpose, like compost or woodchip, needs to be at least 5 centimetres (2 inches) thick and allows us to mulch irregular-shaped areas and go right up to plants, which is not easy using plastic or rigid materials. However, the cost of applying mulch for larger areas means that many commercial growers have tended to use plastic alternatives. Though plastic mulches can work on perennial plants and trees, they tend to be very hard to remove from the system once they begin to break down. The risks of plastic contamination of soil are beginning to be more understood, and since the type and condition of mulch plastic is not easily recycled many growers are now looking for environmentally sympathetic alternatives. Biodegradable plastics have been around for a while. I was using them twenty years ago as a grower, and they have improved in quality since then. They are, however, more effective if you have the scale to invest in specialist laying equipment and mainly have a relatively short life, so they are most suited to crops that only need weed protection for a few months.

In woodchip mulches, the species, size and age of chip will determine how effective and long lasting the mulch is. Smaller chip size is more effective at holding moisture but will break down quicker; similarly, older chips that are already partly broken down will not last as long as fresher material. Imagine, too, how easy it is for windblown seed to germinate on a mulch. Soft, partially composted chip is an ideal surface for seeds to grow on, while larger chips will prevent germination. One is not necessarily better than the other but will be more suited to different situations. For mulching no-dig vegetable beds I would choose a well-rotted fine grade, while for shrubs and trees a larger-sized chip, less broken down, would be more effective.

There is significant variation amongst tree species in how long a mulch will last. In one trial at Royal Horticultural Society Wisley in England, the grower found that hornbeam lasted longer than the alder, holly and compost mulches they compared it with.[2] The hornbeam chip was still providing effective weed control in the second year after application. This could result in reducing woodchip applications from yearly to every other year, which would halve the cost and effort of application. In many cases, it is not possible to source single-species woodchip, but where we can it is worth thinking about the individual characteristics of each species.

WATER RETENTION

The other key use of mulches is to keep water in the soil where it is useful, rather than losing it to the air. It does this by shading the soil and thus lowering the temperature, and by protecting it from the wind, both of which reduce evaporation from the soil surface. Exact reductions in water loss will vary with system, mulch type, soil and climate.

As you would expect, the hotter and drier your climate, the greater the benefit is likely to be.

Plastic mulches give the same, if not greater, water retention but can create other problems. Solid films create an anaerobic environment under the mulch, which can lead to compacted and lifeless soil. Even the woven mulches, which do allow infiltration and some breathability, are not as good as an organic material for healthy soil. In wet climates slugs and snails can proliferate under plastic mulches where they have a safe, moist home away from most predators.

MODERATING SOIL TEMPERATURE

Using a mulch of any sort can help to warm cold soils and cool hot ones. The mulch gives protection to soil and plant roots, shielding it from extremes. It acts as winter coat, sunscreen lotion and umbrella combined.

As with water retention, there have been numerous studies looking at the effects of mulch on temperature. In many situations, plastic mulches will give higher temperature differences than organic materials. However, what I intend to illustrate in this book is that when looking at the multiple benefits a woodchip mulch gives you, it is worth sacrificing a little on water retention and temperature to get the long-term soil health benefits of woodchip mulches.

PHYSICAL DAMAGE

Bare soil is vulnerable soil. Mulching protects soil from physical damage such as heavy rain or wind, and in extreme weather conditions also helps to stop soil being washed or blown away. Woodchip mulches allow water to gently percolate through to the soil, whereas solid plastic mulches prevent the water from reaching the soil, instead causing localised flooding issues as the rain runs off the plastic. Woodchip mulch even helps reduce damage from compaction caused by us walking or driving over the soil. The mulch cushions and absorbs our impact and keeps the soil in good condition underneath.

PEST AND DISEASES

Woodchip mulches have the potential to reduce some pests and diseases. Applied at the right time, they can cover over fungal spores on the soil and prevent them spreading when rain splashes back up onto the plant leaves. The mulch from some tree species will deter pests; for instance, cedar chip can help to keep termites and beetles away. New research in the UK has also been looking at using the salicylic acid in willow woodchip to stimulate an immune response in apples to help prevent scab.

However, we should also be aware of the risk of introducing some pests and diseases when using woodchip. We might be providing a perfect habitat for some insects and other invertebrates like woodlice. Though these are not usually major pests, they can, in some cases, cause crop damage. Similarly, there are tree diseases that might be present in woodchip that could cause problems in some circumstances. These risks can usually be managed but are worth keeping in mind.

PATHS

Woodchip is a great and versatile material to use for mulching paths. Because it is tough and doesn't get damaged by our feet, it will easily last a season of being walked on. As it breaks down, you can either rake up and add to your compost heap or onto growing beds, or just lay another layer on top. The fungi and other organisms in the soil will help to move the nutrients and organic matter around your plot.

Propagation

Though composted organic material is often included in plant propagation mixes, it has not been so common to make use of woodchip for this purpose. There are commercial substrate manufacturers that use bark as the main ingredient for their products, but it is relatively easy to make your own plant-raising substrates from woodchip, as we will find out in chapter 4. We'll also look at using woodchip to make a propagation hotbed for your tender seedlings and as top dressing for potted plants, giving many of the benefits that I have outlined for soil-grown plants.

Mushrooms

If you have ever been into a woods or left an undisturbed pile of logs in the corner of your field, you will know that where there is rotting wood you will find fungi. Most of the mushrooms grown for human consumption live on wood in their natural environment. This gives us a big opportunity to bring mushroom cultivation into our use of woodchip. I am still very much at the beginning of my fungal journey, but I am experimenting with growing mushrooms in bags of wood-chip (which will of course be used again for something else when the mushrooms are done) and on the woodchip mulches in agroforestry systems. I will look in a bit more detail at some simple ways to have a go at growing mushrooms in woodchip in chapter 7.

Woodchip for Animal Bedding

Though not core to this horticultural-focused book, many gardeners will either keep some livestock, even if that is just a few chickens, or have access to farmer friends. It would be remiss of me not to have a brief look at the potential of woodchip as bedding for animals and the added benefit this gives to the subsequent material. Woodchip can soak up a lot of nitrogen and even outperform some traditional bedding materials, aiding animal health and welfare, and reducing leaching both from barns and from the resulting compost. In chapter 2 you'll find more details about the benefits and practicalities of using woodchip for animal bedding.

Soil Health Amendment

It's hardly a news story that adding organic matter to soil is a good idea. Crop rotations with livestock, adding compost and muck spreading on farmland are all based on the principle that for a healthy, productive soil we need to add carbon and nutrients back into the soil to feed the organisms that live there. For some reason, though, we haven't fully exploited the potential of woodchip for this purpose. This may be partly due to availability. Woodchip is a relatively new material; before woodchippers were invented, most woody waste would likely have been burnt, either harnessing the energy for fuel or just as a bonfire to dispose of the material. Tree surgeons now produce millions of tonnes of chip a year, and chippers are readily available to woodland managers and contractors. Some of this chip is a valuable commodity that can be sold to the biofuels industry, but much of it is lower grade, particularly where it has a high proportion of green bark and leaf material, though this might actually enhance its value for our purposes.

There are a few pioneers who recognised some time ago the power that woodchip can add to a growing system; people like Paul Gautschi, of documentary film *Back to Eden* fame, who advocates no-dig gardening, or Gilles Lemieux, whose cutting-edge research in Canada led to the coining of the term 'ramial chipped wood' – both of which are covered in more detail in chapter 5. However, there is now a growing understanding of the role that a range of organisms, but most notably fungi, can play in improving soil health and the part that woodchip has in this process.

There's Something in the Woodchip Pile!

While I am a fully signed-up member of the Woodchip Appreciation Society, there are nevertheless a few risks to be aware of when sourcing and using this material. Most, however, can be easily avoided or mitigated.

Sourcing

Know your supplier. In the absence of recognised standards and certification of woodchip for horticultural purposes, it is important to ask a few basic questions about where the chip comes from. Most crucial is that the chip only contains untreated timber; in other words, that it is made only of trees, shrubs and hedges, and not 'waste' wood, such as pallets, chipboard, garden fencing, etc. Treated wood usually contains chemicals that would leach into your soil and may also contain nails, bolts, staples or manmade coverings, such as plastic or Formica. For this reason, I tend to look to tree surgeons or woodland managers rather than lumber yards for my chip.

Disease

Many trees are cut down and chipped because they are dead or dying. This might have been due simply to old age, but often the tree will have a disease. Even some of the more concerning diseases, however, appear to pose a relatively small risk of spread through woodchip. Honey fungus (*Armillaria mellea*), which is relatively common in older trees, spreads either from local fungal dispersal, through mycelia and root contact, or through infection at propagation. It is unlikely to transfer through woodchip. Some ongoing research in the US, as yet unpublished, on fire blight (*Erwinia amylovora*) showed that the bacteria in infected branches that were cut out, put on the ground and flail mowed became nonviable in a number of weeks. This potentially opens the door for growers to recycle this material rather than remove and burn it. So, even fire blight risk from imported woodchip may also be low.

I have tended to be relatively relaxed on this front, counting on an initial composting of my woodchip to break down harmful pathogens and reasoning that I am using the chip on young healthy plants, or on unrelated species, which makes infection less likely. However, if you are concerned, I recommend longer composting and turning to ensure that every part of the whole pile reaches a temperature that will reduce or eliminate the danger. Linda Chalker-Scott suggests that there is little risk provided that you do not incorporate the woodchip into the soil and that you make sure the chip is not in contact with tree bark where the contact can cause infection in moist conditions.

Coloured Mulches

While most coloured mulches use iron- or carbon-based dyes that in themselves are unlikely to be harmful to soil, there are some that may include toxic colourants. More of a risk is the source of the wood used to make coloured woodchip. Much of it comes from recycled wood such as old pallets and fencing. This wood is favoured for a couple of reasons; it looks less attractive when chipped than untreated timber, but also holds the dyes better than natural woodchip. Chip made from these recycled products can contain the chemicals used to treat the wood such as creosote and CCA (chromated copper arsenate). For this reason I would not recommend using artificially coloured mulches on soil.

Species

Even though we may not get a choice of species from tree surgeons, understanding the composition of our woodchip and knowing how to best manage and use it is still useful. We'll explore the specific properties of different species later in the chapter, but there are a few things to note as general points. Deciduous species generally provide more useful chip than coniferous ones. Though a proportion of coniferous chip is fine for most purposes, the usual recommendation is to include no more than 20 to 30 per cent There are some exceptions, for instance if mulching acid-loving plants like blueberries you could have a higher proportion of conifer chip and might get better weed control as a result. There are also some species that can be dangerous to animals if eaten.

Sawdust

Be very wary of using sawdust. It comes mostly from old dry heart-wood, so it is very high in carbon and low in nitrogen. Its very fine structure also means that it can take a long time to break down, and both water and air struggle to penetrate. Never use it as a mulch or untreated soil additive. Sawdust is still a useful ingredient in an integrated composting system, especially if mixing with a high nitrogen source like animal manure, but it should be used in moderation.

Woodchip from Single Tree Species

While most of us may not be able to source single species of woodchip, I'm fascinated by the individual properties of different species. When planning to plant our own trees as a source of woodchip, this can help to inform tree choice. There are some general principles; for instance, softwoods (as the name suggests) tend to break down more quickly than hardwoods. Mixing them with chip from hardwood species can reduce the need for frequent mulching. In a commercial setting, the labour costs of spreading mulch are high, so this could be a significant saving. There are some species that are poisonous to animals, so if using your woodchip as bedding or in a situation with livestock, check the species and whether it will be suitable for your needs. For instance, there have been cases of cattle and horses dying from eating black locust (*Robinia pseudoacacia*) chip. Rhododendron (all species) and yew (all species) chip are also known to contain compounds toxic to animals. Alternatively, compost the chip long enough for the toxic chemicals to be broken down – usually around six months is sufficient.

Here I have pulled out some of the specific attributes we might want to look at when considering different species. The list is, of course, not exhaustive, and for many species good data on the use of woodchip is not available.

Nitrogen Fixers

Leguminous trees, like their smaller cousins the clovers and beans, can pull nitrogen from the air due to their symbiotic relationship with nitrogen-fixing bacteria that create and colonise nodules in their roots. These bacteria turn atmospheric nitrogen into ammonia, which is a form of nitrogen that is available to plants. The bacteria store the nitrogen in the root nodules. The plants take some of this nitrogen, trading it for carbohydrates that the bacteria cannot manufacture for themselves. When leguminous plants die, or are eaten, the nitrogen that is stored in them becomes available to other plants and animals.

Aside from a small proportion fixed by lightning, this biological process of nitrogen fixation is the main way that nitrogen moves from the atmosphere, where it is plentiful, to the soil, where it is often scarce and needed for plant health and growth. Though there are lots of free-living nitrogen-fixing bacteria in the soil, the combined power achieved by teaming up with plants that can feed them increases the rate of fixation and therefore the fertility in systems where this occurs.

A 2019 modelling of tropical forestry growth found that forests with nitrogen-fixing trees significantly increased their growth and ability to sequester carbon, potentially doubling the carbon stored in the first thirty years of their life, and even at maturity giving 10 per cent greater carbon storage than forests without nitrogen-fixing species.[3]

Leguminous trees tend to be fast growing and therefore useful as pioneer species in new plantations, giving protection to main species. Once they have served this purpose, they can potentially be removed (or coppiced) and chipped.

ALDER

There are approximately thirty species of *Alnus*. While their nitrogen-fixing potential will vary between species, as an example, red alder (*Alnus rubra*) can sequester as much as 300 kilograms (660 pounds) of nitrogen per hectare per year.[4]

I am involved in an experimental agroforestry project at Eastbrook Farm in Swindon in England, where we are planting a range of trees in a silvopastoral system. Silvopasture is the deliberate integration of trees and livestock in a farming system, which brings a whole range of benefits to the animals, soil and wider environment. We have included European (*Alnus glutinosa*) and Italian (*Alnus cordata*) alder in some areas, with a view to boosting the overall productivity of the system, but also to coppice and use the chip. A study by the US grants programme Sustainable Agriculture Research and Education showed an increase in soil organic matter from 10 to 11, 4 per cent from adding alder woodchip as a mulch for three years.[5] Alder could also be classed in the quick-growing category on page 14 and is very useful as a nurse tree for other trees in windy situations.

BLACK LOCUST

I became aware of the potential of *Robinia pseudoacacia* when the American silvopastoralist Steve Gabriel visited the UK in 2019. Though sometimes invasive, if properly managed it seems to have massive potential to provide nitrogen and plentiful woodchip. It is traditionally used for fencing posts because it is resistant to rot, which suggests it may have some potential as a species for woodchip used as a weed control.

Be careful when using fresh chip from this species as it can be poisonous to some livestock. I aim to use it in areas without stock or will compost the chip well, before use. The ferocious thorns on it may cause a problem both during the chipping process, which can by all accounts be a bloody process requiring full body armour, and to vehicle tires if material is left anywhere you will be driving.

Allelopathy

The word derives from the Greek *allelo* (mutual) and *pathos* (suffering). Though it can have a beneficial effect, allelopathy usually describes the ability of one plant to release a chemical that has a detrimental impact on other plants or organisms, thus preventing competing species from germinating or growing.

There is debate around allelopathy, and in particular the difficulty in distinguishing between the effect of the supposed allelopathic chemicals and other competitive tendencies between plants; for instance, taking light and moisture from less robust neighbours. Whatever the exact cause, some species do seem to have an ability to prevent competitors from thriving near them. Here are a couple of examples.

WALNUTS (AND OTHER MEMBERS OF *JUGLANDACEAE* FAMILY SUCH AS HICKORIES AND PECAN)

Walnuts are perhaps the most famous example of an allelopathic tree. Walnut trees produce juglone, which is a chemical said to harm some other species (look out for symptoms like yellowing leaves, wilting and even death). Juglone is present in all parts of the tree, including the wood and fruit. The black walnut (*Juglans nigra*) has the highest concentration of this chemical.

Is juglone really the culprit though? R.J. Willis has shown that there is no definite proof that roots can take up juglone, so we should be cautious about attributing the competitiveness of walnuts just to allelopathy.[6] However, there is something going on; the juglone does have an effect on the life in the soil. Willis states that: 'Wood and bark particles from trees containing juglone … are reported to decompose much more slowly than those from other hardwood species and earthworms have been recorded as being scarce in walnut areas.' Summers and Lussenhop found that cores of walnut soil, when removed beyond the influence of the walnut canopy, increased in both arthropod number and diversity. Moreover, the authors concluded that: 'The greater amount of soil organic matter associated with walnut could be ascribed to the general retardation of decomposer organisms.'[7]

Some plants, like asparagus and nightshade, are particularly susceptible to competition with walnut, though whether this is attributable to juglone is not clear. European alder, some birches, crab apples and Norway spruce can also be affected. The good news is that even if juglone is to blame, it breaks down quickly and six months of composting woodchip from black walnut is enough to reduce the levels to be safe even to the species at most risk.

EUCALYPTUS

The allelopathic potential of eucalyptus is stronger even than black walnut. The distinctive-smelling oil contained in the leaves is very rich in terpenoids, which can have a wide range of biological actions including antimicrobial, fungicidal, insecticidal (and insect repellent) and nematocidal. It also has strong herbicidal properties. A whole range of products for plant, animal and human health have evolved from these properties. But what may work in a concentrated product is not necessarily replicated from using the chip. Most of the studies I have seen seem to show that woodchip from eucalyptus does not have any detrimental effects on soil or nearby plants, and indeed provides a long-lasting mulch. It appears that the majority of the allelopathic effect comes from the leaves, which contain a higher concentration of oils than the woodchip does.

Many woodchip animal bedding products use eucalyptus as part or all of the mix, due to the antimicrobial benefits. The trees themselves are well known for repelling flies, thus providing a welcome refuge for livestock during hot weather. Most eucalyptus are native of Australia and therefore not entirely happy in colder climates. Hard frosts or long periods of freezing temperatures can damage or kill the trees. But if your climate suits, you keep animals and were planting your own trees to provide woodchip, it might be worth including a few eucalyptus trees for fly repellent purposes, knowing that the chip from this very fast-growing species would provide a useful resource as well.

Quick Growers for Coppicing

While high potential for growth and yield may not be the primary concern for choosing a species, I am certainly interested in how much chip I might get from different trees. It might also have a bearing on price if I am buying it in. Quick-growing species are often turned into

Short rotation hazel coppice in an agroforestry system at Wakelyns Farm.

woodchip, partly because they can be coppiced rotationally, making harvest easier, but also because they are less suited to other uses than harder, slower-growing species. This also means that they are quicker to rot down, but this may not be a problem. Indeed, in some circumstances it could be an advantage, for instance in producing ramial chipped wood for improving soil health (see page 109).

Growing your own short rotation coppice is a great way to provide not only woodchip, but also shade and shelter for crops and animals, as well as additional forage and habitat for wildlife. In some circumstances, like growing willow on land that periodically floods, it may be an opportunity to turn unproductive ground into a valuable resource.

WILLOW

Salix is the classic quick-growing tree. New hybrid varieties are so productive they can produce more than 40 tonnes of dry wood a year per hectare, though these may not be the most suitable varieties for

Short rotation willow coppice in an agroforestry system at Wakelyns Farm.

our purposes. Willow is popular for other reasons; it provides good browsing for livestock and grows well in wet areas that might not be suitable for other crops. It also has a high salicylic acid content; this is the substance from which we get aspirin. There is much anecdotal evidence that livestock will selectively browse willow to self-medicate and, as we shall examine later, willow might even have specific health benefits for trees when used as mulch. Willow is also cheap and easy to establish and is therefore a natural choice for those of us looking to establish large areas of trees.

POPLAR

The fast-maturing *Populus* is widely used as a windbreak species, though it also performs well in a coppice system. It will initially grow a single thicker stem than the willow, meaning that the first cut will produce larger logs. Overall, poplar is slightly less productive than willow, but planted in combination would improve biodiversity since a different range of moths and other insects make use of willow and poplar. Like willow, poplar has high levels of salicylic acid.

HAZEL

Corylus develops more slowly than willow or poplar but produces harder woodchip. If grown on a longer rotation you might be able to get multiple outputs such as nuts, or other products such as poles for hurdle making or garden stakes.

CHESTNUT

Castanea sativa is one of the traditional coppiced species when grown for building and fencing. Though some other species can also be used the fast, upright growth of this species makes it particularly suitable. Combining this output with using the offcuts for chip could be useful, or even on a longer rotation devising a system that allows for the harvest of nuts, though I don't know of anyone doing this on a commercial basis. In the UK, our climate does not yet produce consistent commercial nut yield from chestnut, though with climate change this might not always be the case. Chestnut is a slower-growing species, so the annual yield of the chip would be lower; however, as with hazel, the chip is harder and would therefore last longer.

Coniferous (Softwood) Species

This covers a range of species that might have different individual properties. One thing they have in common is that their bark and needles all have a low pH. Is this a problem? Though we think of material from coniferous species as being acidic, we can see in Table 1.1 that even the wood from broadleaved trees has a low pH before being chipped. There is also some difference between the heartwood and the sapwood – the pH of the latter normally being a little higher.[8] However, as soon as it is chipped and begins to break down, the pH rises, as organic acids break down. This is true of coniferous species, too. Secondly, the amount of woodchip you are applying (even when mulching thickly) is relatively very small compared to your soil, and evidence suggests that it is almost impossible to change the pH of your soil by adding organic amendments. Even in controlled environments, adding composted pine bark at 20 per cent in a potting mix had no detrimental effect on the plants.[9] Similarly, in a three-year experiment with pecan tree woodchip there was no effect on soil pH, though soil organic matter increased.[10]

There are some potential challenges with composting woodchip that has a high level of coniferous material, in that the initial stages of decomposition are slower. From my own experience, in concentrations

Table 1.1. Comparison of pH of Woodchip from Various Tree Species

Hardwoods		Softwoods	
Species	pH of Woodchip	Species	pH of Woodchip
Elm	6.0–7.2	Sitka spruce	4–5.5
Beech	4.5–5.9	Douglas fir	3.75
Chestnut	3.6	European larch	4
Painted maple	4.75	Western red cedar	2.9–4
Black walnut	4.7	Lawson cypress	4.35
Oak – European and US	3.3–3.9	Scots pine	4.3–4.6

of up to about 30 per cent in a mixed woodchip batch there should be no issues at all. In higher concentrations, the composting process may be slower but the resulting compost will still be a great product. If used as a mulch, the release of weed-inhibiting chemicals from many coniferous species may even give a short-term advantage.

CHAPTER TWO

Sources of Woodchip

Trees grow in most places in the world. So, woodchip should be a readily available material. There are still choices and decisions to make, though, in where you look for a good supply. Quality, cost and distance are just some of the factors to consider if you are getting it from someone else. While those with their own land and trees must decide what species to plant, how to manage them, and what methods to use for harvesting and chipping. Sustainability and the environmental impact of farming trees for woodchip on a commercial scale is also an important consideration. In this chapter, I will go through most of the options for sourcing woodchip, looking out for pitfalls along the way.

Woodland

This chapter does not pretend to be a guide on how to manage woodland, but rather it aims to highlight the opportunities for obtaining woodchip from these widely underutilised areas of trees. In most countries, there are areas of financially low-value wood available to those with enough motivation and knowledge. Even in the UK, with our relatively high population density, there are a surprisingly large number of undermanaged woodlands. Frequently, these areas are ancient woodlands or were planted without a clear market or have been neglected post-planting. The result is often that the trees have little commercial value for higher-grade timber, or, even where they might be good enough, are available in such small volume that most lumber yards are not interested. Though there are a few small mills that look out for interesting wood in small quantities, most of the wood and coppice arising from these neglected woodlands are only suitable for firewood or chipping.

Managing woodland well can help not only increase the potential income from that land, but also increase biodiversity, enhance communities and create jobs. Before piling in with a shredder and removing all the wood for chip, though, we should do a proper assessment of the trees and the biodiversity. We need to be looking at establishing long-term (twenty-plus year), medium-term (twenty year) and short-term (five year) objectives. This includes identifying the following:

- Trees we want to keep; for instance, for biodiversity value, to provide a long-term canopy or in a grazed woodland setting for browsing.
- Trees that might have some future timber value. This could include as logs for firewood as well has higher-grade lumber.
- Trees and underwood we might want to use for woodchip.

For some areas of ancient woodland, it might be more appropriate to leave them undisturbed rather than actively manage them. Others, such as ancient coppiced woodlands, will have a history and evolved ecology based on that cycle of cutting and would benefit from a return to that management. The long-term health and sustainability of any area of woodland is also important. In recent history, there has been an approach of creating a 'tidy' woods by burning or clearing any fallen branches or brush, which I remember seeing twenty-five years ago in the UK. The impact of historical management and exploitation is thankfully now being replaced by a greater understanding of the value of decaying woody material to the well-being of the woodland. However, this also changes the judgement of what wood is 'waste' and what, if any, is valued. As with any resource, there is an appropriate and sustainable level of exploitation, and we should aim to only harvest wood for chip that will not have a significant impact on the wider environment from which we are taking it.

As an example, I recently visited a new horticultural enterprise interested in using woodchip as part of a small commercial no-dig system. The estate of which it was a part had a few hundred acres of trees, but the forester was implementing a long-term plan to revitalise the woodland and so was reluctant to part with material for chipping that he felt was essential to restoring the biodiversity and fettle of his woods. In the end, the grower decided instead to plant some short rotation coppice across the market garden to provide their own woodchip and bean poles. By carefully siting these trees they also got wind protection for their crops and increased habitat for beneficial wildlife.

If we own the woodland ourselves it is relatively straightforward to assess the potential and costs, and take advantage of the resource. However, if we are approaching other woodland owners with a proposal, we need to sell them the benefits of what we are doing. This will depend on what stage their trees have reached but could include harvesting material from ride edges and glades to keep them open. Other options are non-commercial conservation coppice management to encourage butterflies, or taking smaller material not suitable for the firewood or timber markets. Keeping rides open and in sun is not only good for insects and birds, but also prolongs the life of the roading by keeping it dry and free of leaf fall material.

The first thinning of younger woodlands can end up costing the woodland owner, since the value of the extracted material is lower than the cost of the operation. It still needs doing, though, or the remaining trees won't otherwise be able to develop. We can offer to cut these early thinnings in return for the wood, or we might even be able to charge a small fee to the landowner.

What about more established woodlands? In theory older woods will have more wood as the trees will be bigger and should therefore offer more commercial opportunity, not just for our increased volume of woodchip but for a range of other markets. But bigger trees need bigger chippers and access is likely to be an issue. This adds to the cost and, therefore, the viability. There is also more risk of damage to the understory and soil.

Farmers and foresters inhabit different worlds, and rarely talk to each other. Though agroforestry systems are helping to bridge this divide, this lack of mutual understanding is the cause of a multitude of missed opportunities. An ignorance of the timber market has led some farmers to feel taken advantage of when working with foresters. If you don't know the value of your timber, it is easy to be exploited. Similarly making the most of your trees is tricky if you don't know how to manage them or what specification you are aiming for. For this reason, many farmers have ended up just ignoring their woodlands

A mix of coppiced and taller trees growing in a shelterbelt. The range of heights and management gives good shelter and biodiversity benefits.

and not seeing them as a productive part of their farm. Offering a fair and transparent profit share deal that gives them a reasonable benefit or return is an opportunity to bring some of these areas back into active management.

However, even when there is little financial incentive to the landowner for selling wood or saving management costs, there may still be opportunities to obtain woodchip. You may need to be prepared to do the work for nothing in return for the woodchip you extract. Altruistic landowners may be swayed by the biodiversity improvements or better access to their woodland, for instance, to allow you to gather material for woodchip.

If you are taking on the long-term management of a woodland, as opposed to a one-off thinning or clearing job, you will need to implement a silvicultural system. This is the process of tending, harvesting and replacing trees in the woodland. Your choice of system depends on what is already there, both in terms of species and systems, and on what you want to produce. High-canopy, predominantly single-stemmed trees are known as 'high forest', as opposed to scrub or coppice. The former will hopefully produce timber trees, while the multiple-stemmed smaller trees can be far more readily used for woodchip.

Who Does the Work?

Tree work is skilled and dangerous. It is vital that operatives have the right training and insurance, particularly if you start working on other folks' land. For this reason, many people opt to contract an existing arboricultural business to undertake the chipping operations rather than tackle it themselves, with all the associated costs of upskilling and legal compliance. However, there are other options. In Europe there are great examples of cooperative models and machinery rings. In France there is a group of fifty farms working together to manage 176 kilometres of hedges to produce woodchip for on-farm and other local biomass boilers.[1] The success of this cooperative project has led farmers to plant more hedges as they have been able to see the benefits they can bring financially to the farm.

In Scandinavia, where there are fewer people and more trees, forestry cooperatives are common. Some of them offer their members the whole range of forestry services, from creating management plans right through to harvesting and processing. Much of the thinnings from forestry plantings in Scandinavia are processed through community heating projects.

In Wales, Coed Cymru (the Welsh woodland organisation) have done a lot of work looking at the practicalities of managing farm woodland. One project looked at the woodchip service chain and worked with contractors with a large chipper to offer services to farmers to manage their woodland. The chip was either used on-farm or taken away and sold. Within a group of farmers, there is often one that has sufficient interest and need for woodchip on their own farm to invest in the necessary machinery and training, and who then contracts out their kit and time to the other farmers within the area.

In some areas there may even be a public or community group that could work with contractors to manage woodlands. On Hawkesbury Common in Gloucestershire in England there is a community wood fuel group, where residents within a defined geographical area have the right to manage areas of woodland in return for firewood. The process is carefully managed by Gloucestershire Wildlife Trust that owns the land to ensure the right trees are harvested in a rotation. There is no reason why this sort of model could not be adapted to provide a useful source of woodchip, too.

Managing Hedges for Woodchip

We have a lot of hedges (or hedgerows) in the UK. Historically, it was one way to divide fields to keep livestock in or out. Though we lost a lot of them in the period after the Second World War, as we embraced larger-scale arable production, this trend is now reversed with increasing interest in agroforestry and financial incentives to replant. Some areas of the UK less suited to arable production, such as Wales and South West England, have retained their network of smaller fields and hedges. In the US many boundaries were created at a time when wire was cheaper and more available. This means there are fewer hedges in the US, but, again, there is growing enthusiasm for incorporating shelterbelts and boundary plantings as part of an agroforestry system. Whether you are looking to make better use of existing hedges on your farm, or considering planting new ones, this section will consider the potential of hedgerows to provide woodchip, and some of the practicalities of managing and harvesting that material.

Methods of Managing Hedges

Working out what you want from your hedge will help you decide how you manage it. For instance, is it primarily to keep animals in or out? Is woodchip your main output? Is biodiversity important? The final management strategy will also have to take account of species mix, access and location, and availability of machinery. Maximising production for woodchip will often mean changing how you maintain your hedges.

TRIMMING OR FLAILING

Most hedges in the UK are treated this way. For hedges along our narrow country roads an annual haircut is a necessity to allow traffic access. However, even in open fields yearly flailing has become the norm. The two main drivers for this destructive behaviour appear to be firstly the belief that every inch of field is needed for production and that if you allow the hedge to encroach you lose profit. The second and more deep-seated conviction, and one which is probably harder to change, is the farmers' yearning for 'tidiness'. A wild hedge is a messy hedge, and most farmers still believe that neatness automatically means well managed. We certainly remain paranoid that our neighbours will judge our farming ability on the straightness of our lines and the neatness of our fields. Weeds and hedges alike suffer in our determination to instil order onto those unruly plants that insist on growing in a mess. The neat box edges of a recently trimmed hedge can be looked back on as a job well done, nature tamed, the farm being properly managed.

This approach unfortunately ignores the benefits that a healthy biodiverse hedge with mixed ages of wood can bring, not just with increased wildlife habitats and carbon sequestration but also the profitability of the farm. Increased pollinator and predator populations from healthy hedges can increase pollination and reduce pesticide need.[2] Even moving from a once a year cutting regime to every three years can have a dramatic effect and reduce hedge management costs.

The challenge for harvesting woodchip in this system is that you will need to have a collection system attached to the flail. Though there are companies developing this equipment, it is not yet widely available. Mostly the chips are left along the hedge where they fall. This will of course have some benefit to the soil next to the hedge and may help with crop health of those plants nearest the hedge. Farmers sometimes observe the best yields from the plants along the edge of a hedge, though tying down the reason is not straightforward; shade, shelter, organic matter, pollination and predation, for instance, may all play a role.

LAYING

Laying is the traditional art of partially cutting trees close to their base so that the stem can be bent over and laid almost horizontally; the laid over stem must remain attached to the base with enough bark and sapwood to keep the branches alive. The laid branches are then woven together to form an impenetrable barrier. Before the invention of barbed wire, using thorny species in this way was the main method farmers used to keep their livestock where they wanted them. A laid hedge is still the best way to provide the densest barrier. A well-laid hedge is a beautiful thing, and in Europe there are distinct regional styles of doing it. More than thirty styles are recorded just in the UK, with more in France, Germany and Holland. The hedge will grow out of the laying and the process needs to be repeated approximately every ten to fifteen years. In the interim the hedge is often trimmed or browsed to keep it as a hedge. There is no reason that a laid hedge cannot be a source of woodchip; indeed, there is usually a quantity of material that is produced during the process that will need to be dealt with. Deciding whether to lay a hedge will not depend on your plans to produce woodchip, but doing so is not a barrier to producing plenty of useful woody output. Through-out most of history, hedges were the main source of fuel for farmers, while the coppicing produced useful poles.

COPPICING

Hedges are often coppiced, or periodically cut to ground level, as part of a long-term strategy; for instance, if they have become patchy or ragged and you want to prepare them for laying. Coppicing first and allowing some years of regrowth gives good straight stems suitable for hedge laying. However, a planned ten- to fifteen-year coppicing plan can provide a productive hedgerow, with relatively little management, that is also fantastic for wildlife. Once your rotation is established you can have some hedges at all stages of development. The frequency of coppicing will depend on species of hedge, speed of growth in your location and use of material. For example, if you want more logs for firewood from the hedge you would leave longer between cuts to give

Top binding a Midland-style hedge, National Hedgelaying Champion-ships, Leicestershire, England, October 2007. Photo courtesy of Hedgelink and Robert Wolton.

the stems time to thicken. Repeated coppicing will change the nature of a hedge and leaves a gap for a year or two where there is no stock barrier, so it is not suitable in all situations.

Pollarding, where the growth is cut to a point on the stem, usually a couple of metres (about 6 feet) above ground level, can similarly produce a range of sized wood. In some systems a thick, stock-proof hedge can be left at the lower level, while individual trees are pollarded at intervals to provide woodchip.

Rescuing a Neglected Hedge

Many hedgerows are scorned and underused. So here is a little guidance on how to bring back such a hedge to productivity. For more comprehensive information on this, check out the excellent booklet 'Productive Hedges' from the Organic Research Centre.[3] The first step they recommend is a full assessment of all your hedgerows; this will help identify those that are suitable for different management options. For

A neglected hedge being brought back into use by pollarding willow.

Collecting wood for chipping using the low-tech method at Waterland Organics, Cambridge, England. Courtesy of Helen Holmes.

example, you may not be able to readily coppice a hedge right next to a road. It will also allow you to tie in hedge restoration with other farm activities, such as replacing wire fences. One of the major tasks if you are starting to coppice a hedge that was not previously managed in this way is removing all the old wire fencing that is likely to now be buried deeply in the hedge, often even grown into the trees. Factor in the time and cost of doing this. It is often easier to flail the sides of the hedge prior to coppicing. This will reduce the material you are working with and make it simpler to see what there is in the middle of the hedge.

Once you've prepared the hedge for coppicing, you're ready to start cutting. Chainsaws are cheap to buy and the most versatile method, and avoid the compaction you can get with larger machinery. However, they are slow, so labour costs soon mount up. For hedge lengths of less than 100 metres (330 feet) the cost of bringing in expensive machinery means that the chainsaw is still usually the most economical choice. There are a range of larger cutting options available like circular saws, tree shears and felling heads. They can deal with large amounts of material quickly, with each having their own advantages. If you have a lot of wire embedded in the hedge, a chainsaw may be your only option to avoid damaging the less precise larger kit.

Costings for Hedge Management

Farmer Ross Dickinson recorded times and costs for a 220 metre (720 foot) long, 6.5 metre (21 foot) high, fifteen-year-old hedge, and this is what the economics looked like. He believes it can be profitable provided there is a local demand for logs.

Table 2.1. Ross Dickinson's Pricing Comparison for Flailed and Coppiced Hedge Management

Operation (For 220 Metre / 720 Foot Hedge)	Cost
Initial flail: 2 hours at £30/hr	£60.00
Manual coppicing: 88.5 hours at £15/hr	£1,327.50
Processing with branch logger: 20 hours at £30/hr	£600.00
Abstraction of nets: 8 hours at £12/hr	£96.00
Abstraction of cord wood: 6 hours at £30/hr	£180.00
Brash burning: 5 hours at £25/hr	£125.00
Processing saleable material and ugly sticks	£750.00
Delivery cost: 15 tonnes at £16/t	£240.00
Total Cost	**£3,378.50**
Sales and Savings (For 220 Metre / 720 Foot Hedge)	**Income**
Savings in annual flailing over 15 years: 3,300 metres at £0.35/m	£1,155.00
263 × 15 kilogram nets kindling twigs = 3.95 tonnes at £190/t	£749.50
99 × 15 kilogram nets of cobs = 2.48 tonnes at £190/t	£470.50
6 tonnes of ugly sticks at £150/t	£900.00
9 tonnes of saleable logs at £181/t	£1,633.50
Total Income	**£4,908.50**
Profit	**£1,530.00**

However, Ross has not made use of the smaller stuff for woodchip. If we strip out all the costs and income from his model relating to the smaller materials we get a budget that looks like Table 2.2.

This would suggest that with the savings made from annual flailing, and income from selling the best firewood, producing woodchip becomes cost neutral. There would, of course, be a processing cost for the woodchip, but you should nevertheless still be able to provide a very cheap chipped material. Ross uses volunteer labour to harvest

Table 2.2. Adjusted Budget for Hedge Management with a Greater Focus on Woodchip

Operation (For 220 Metre / 720 Foot Hedge)	Cost
Initial flail: 2 hours at £30/hr	£60.00
Manual coppicing: 88.5 hours at £15/hr	£1,327.50
Processing with branch logger: 20 hours at £30/hr	£600.00
Abstraction of cord wood: 6 hours at £30/hr	£180.00
Processing saleable material	£450.00
Delivery cost: 9 tonnes at £16/t	£144.00
Total Cost	**£2,761.50**
Sales and Savings (For 220 Metre / 720 Foot Hedge)	**Income**
Savings in annual flailing over 15 years: 3,300 metres at £0.35/m	£1,155.00
9 tonnes of saleable logs at £181/t	£1,633.50
Total Income	**£2,788.50**
Profit	**£27.00**

and process what he calls the 'ugly sticks', which he then sells at cost to people in fuel poverty. This material could also be added to the wood-chip volume, as there is likely to be little profit otherwise. There is a market for the chip from hedges as biomass fuel, though the quality is not always good enough and it will need to be dried to be marketable. For this reason, the economics seem to stack up best if using the chip on-farm. The biggest financial incentive from moving from yearly cutting to a longer-term rotational cut is the saving in trimming costs. In 2015, the Organic Research Centre estimated the annual cost of flailing hedges at a higher figure of £0.35 per metre, which over fifteen years adds up to £5.25 per metre.[4] They predict that: 'For a farm with 10 miles or 16.1km of hedges, where half are managed by coppicing for wood fuel and 400m are coppiced every year, £29,880 could be saved in reduced flailing costs over 15 years, not including the potential cost savings from using the woodchip as fuel or the income generated from the sale of the woodchip.'[5]

Estimating yield of woodchip from hedgerows is difficult as there is so much variation between hedges. The Organic Research Centre measured the output from the first cuts of a number of hedgerows, showing a range from 21 cubic metres (740 cubic feet) to 35 cubic metres (1,235 cubic feet) of woodchip per 100 metres (330 feet) of hedgerow.[6]

Machinery for Hedgerow Woodchip

There is more information about machinery in chapter 3; however, there's more than simply cost and availability to consider when looking at which system of hedge management you choose. Access can be tricky for hedges. They may be next to a road or only be reachable through a narrow ingress. As we have discussed, the presence of old posts and wire within a hedge may necessitate manual chainsaw work. Though there is talk of machinery that will harvest the chip directly as it is flailed or cut, I have been unable to find anyone producing or using such a piece of equipment. In the meantime, the options for harvesting woodchip from hedges are limited to when they are laid or coppiced.

Agroforestry

Agroforestry is a system of farming or growing that deliberately integrates trees with other farm enterprises. Rather than having a separate orchard or woodland, trees are planted within fields either as rows, clusters or individuals. The interactions with the livestock and crops bring a multitude of benefits both to the tree, animal and other plants, as well as to the wider environment.[7] If you trace our domesticated flocks and herds back to their wild ancestors, we would find them grazing or foraging amongst trees, in or at the edges of woodlands, in a landscape of wooded pasture where groups of trees were interspersed with clearings and grassland. Animal welfare and productivity generally increase with trees. They give shade and shelter and, for many animals, an additional food source. There are a range of similar benefits for crops. Temperatures in a tree-sheltered area can be up to 10 degrees Celsius higher in cold periods, though is usually more likely 2 to 4 degrees Celsius more.[8] Growers can exploit this for a longer growing season, quicker growth and potentially even a wider range of crops that can be grown. The biodiversity that those trees bring also increases predator and pollinator numbers, which can have a positive effect on plant health and reduce the use of pesticides. Integrating trees can improve soil health and organic matter, help with rain infiltration to keep soil drier and mitigate the risk of flooding in wet periods, while aiding moisture retention in drier times. The trees help to make the best use of deeper soil, light and water to increase the overall productivity of your patch without any additional inputs.

For those looking to implement an agroforestry system, the choice of tree and system can be a bit overwhelming. There are so many species you could grow and an almost infinite number of planting arrangements. Focus on the function of the trees. What is the main output you are looking for? Fruit or nuts, timber, woodchip or shelter for livestock? Trees often provide multiple benefits, but assessing each possible species according to the main use is a good way to narrow it down. Choosing a planting system often comes down more to how you want to manage or harvest the crop. Straight lines make management easier and add efficiency on larger plots, but blocks or informal groupings might be better for livestock.

Trained fruit trees are a perfect option for domestic and market gardens. They can be pruned to almost any shape or size. For very small spaces, stepover-trained trees, i.e. a single horizontal stem running at about knee height, can be run alongside a path or between rotations with very little competition to its neighbouring plants. Espaliered or fan forms do a similar job for bigger gardens, and, if kept to 2 to 2.5 metres (6 to 8 feet)

Pollarded willow with a chip pile in front of a hedgerow at Eastbrook Farm.

high, should be manageable and allow good growth of interplanted crops. However, trained trees take time to maintain; commercially this time is only justified for high-value fruit crops. For large farms, growing big trees for timber can work well, though of course they will have an impact on the land around them, and eventually it will not be possible to grow other things nearby, though animals will still benefit from their shade and shelter.

Cutting entire trees down or lopping their branches on a regular cycle is another, simpler method to control their size and get an ongoing output. Coppicing is where you cut the tree to ground level, while pollarding leaves a section of trunk, usually just above the height at which animals can browse, and the branches are cut to this point. The aim is to produce regrowth for a range of things like wood fuel, bean poles, fence posts, basket weaving, tool handles and thatching spars, and leaf browse to name a few. However, it is also a perfect system for producing woodchip, especially ramial chipped wood (see page 109). The technique produces straight, thinner stems that are efficient to harvest and fit well through a chipper. Most deciduous species respond well to coppicing and pollarding, but some like willow, poplar, hazel, lime and chestnut are especially suitable.

Woodchip Yields from Coppicing

There is a lot of data on yields from high-performing willow hybrids that have been bred for biomass boilers. Even poorly achieving sites can get 4 ODT (oven dried tonnes) per hectare per annum, while the very highest output can be more than 30 ODT per hectare. Species or hybrid choice and site can make a difference, but a 2010 review showed that 10 ODT per hectare a year was a reasonable expectation.[9] Poplar will give a broadly similar yield, though won't hit those high yields of the best willow. Once you move away from the willow and poplar, data on yields is a little harder to come by, but if we pull in some information from the paper industry, we can make some broad comparison of a few species.

As we will see in chapter 5, the Organic Research Centre also did some work on yield per tree as part of their ramial woodchip research, showing some examples of cubic metre output from a hedgerow situation. While none of these figures should be used

Table 2.3. Estimated Yield of Woodchip from Sample Tree Species

Species	Yield (ODT per Hectare)	Stems per Hectare
Willow	4–30	15,000
Hazel	2	1,500
Poplar	4–20	9,000
Alder	16	15,000
Eucalyptus	16–22	1,500
Softwoods	7–20	2,500

as an accurate predictor of yield, they help to give you an idea of how many plants you might need for each tonne of woodchip per year produced. It is also worth noting that trees planted in single or double rows will perform differently to those grown in a dense coppice or woodland situation. They will have less competition for light and moisture, which could help them grow more quickly. However, they will be more exposed, which could have a detrimental effect. It has also been shown that diverse mixtures of trees (as with most plants) are more productive than monocultures.[10] This suggests that, although it might make it more complicated to harvest, from a woodchip yield point of view, combining species even within a row of trees might be worthwhile.

System Setup

Since there are so many variables when designing an agroforestry system, by necessity these guidelines are somewhat general. The two key starting points are:

1. Do a full assessment of what trees you already have and how you might make better use of them, or whether they can be easily augmented. For instance, are there existing hedgerows that could be widened and managed differently to provide a useful output.
2. Identify your key outputs from planting trees. In this case, I am assuming that woodchip is the main driver, but it is likely that you'll have other benefits or products you are looking for: logs for fuel, shelter for crops or to feed animals, for example.

Having done that appraisal of existing trees and what you are looking for as a harvest, there are a few options for planting up an agroforestry system, though they can of course be interlinked and mixed.

USING EXISTING TREES AND HEDGES

We've already looked at how a change to hedgerow management can give you more usable wood material. A similar approach should be applied to other woody assets. Are there clumps or lines of trees that could provide woodchip if maintained differently? Thinning or coppicing some trees within a group could help biodiversity and the health of other trees within that stand of plants, as well as giving you some chip. If you are unsure of the current ecological value of your trees, consider bringing in an ecologist before wading in and chopping down trees. You should be looking for a sustainable approach to managing the habitat that has an overall positive impact on the environment. Many hedges will need gapping up, or enhancing, and you might consider increasing the widths of the hedge, not just by letting it grow bigger, but also with an extra row of planting. Double-rowed shelterbelts can be more effective on windy sites provided there is still wind porosity.

SHELTERBELTS

These can be thick hedges or might be wider lines of trees, often along riverbanks and field boundaries. Designing shelterbelts for maximum benefit is a scientific art. A solid block of evergreens creates eddies on the sheltered side as the wind hits the barrier and flows over the top. A permeable barrier is of more use as it allows some air through and acts to slow rather than stop the gales. This usually means growing mostly deciduous trees to allow wind to filter through, especially in winter when the winds will be harsher. The angle at which the wind hits the trees, and the length and height of the belt compared to the area you are looking to protect, will affect how well the shelterbelt works, too.

As shelterbelts get older there is a tendency for lower branches to die off leaving the bottom open, meaning wind races through underneath the higher tree canopy. If you have planted the shelterbelt for lower-level protection this effectively makes them useless. Selectively thinning or coppicing sections of your shelterbelt will provide you with woodchip and will help to maintain the mixed permeability needed for protection.[11]

GRAZED WOODLANDS AND INDIVIDUAL TREES

Though not typical systems for woodchip production, individual trees and grazed woodlands could make a contribution. They tend to be widely spaced larger trees, which provide shade and shelter and some browse from lower branches. However, as with any woodland area there will be need for occasional felling, which could give some wood for chipping, particularly in the development phase. You might consider managing some trees within the woodland by coppicing to maintain some lower growth. This material can be chipped or fed to the livestock, or even both if the felled branches are left on the ground to be eaten and the residue woody material is then chipped. It is worth noting that while fine fresh or dried, the wilted green leaves of some species, such as cherries and plums, can be highly toxic. The wilting can be caused by frost or cutting and transforms glucoside in the leaves to hydrocyanic acid and sugar, making them both more palatable and more deadly.

BROWSING BLOCKS

These are areas of densely planted quick-growing species such as willow, poplar or mulberry. They are designed to allow livestock to forage the tree leaves for a short intensive period, then be moved out before they do lasting damage to the tree. For them to work effectively, a balance between growth and browsing needs to be tightly managed. Inevitably, there will be times when this doesn't quite happen. The trees chosen for browsing in these systems are broadly similar to the species you might plant for short rotation coppicing for woodchip, and so could be easily harvested and chipped if they started growing faster than the livestock needed, or if the trees needed rejuvenating.

ALLEYS

Growing single lines of trees across a field, or double rows with one row harvested in any year to give continuity of habitat and shelter, is an increasingly popular system for agroforestry. It suits large-scale mechanised operations and allows easy management of both the trees and other crops or livestock. It is particularly suited to wood chipping as you have easy access along the row with any machinery you might need for cutting and chipping. If you are intending to use the

Case Study

Al and Aurore Whitworth are experimenting using woodchip from coppice around their highland croft in Scotland.

'We moved to our small 8-acre croft on the north coast of Scotland in 2018 with the intention of growing not only food for ourselves, but also all of our firewood. To satisfy our cooking, heating and hot water we get through a lot of wood. We are setting up large areas of coppice for this, mainly alder and willow, and will have a lot of small-diameter wood that would be perfect for ramial chipped wood.

'We have so far experimented using chips on raised beds for annual vegetables, some ramial and some bigger chips delivered by a friend. Though too early for definitive results from these experiments, we have seen huge flushes of fungi, as we set up the beds and improve the soil. We intend to continue using ramial woodchip on the beds to bring and maintain fertility, as well as mulching with seaweed. It is important to us to minimise external inputs to the land and so using chipped wood grown here makes perfect sense.

'We are also setting up a small tree nursery, growing native trees in cells from locally gathered seed. This, combined with our annual vegetable propagation, uses quite a bit of compost. This year we hope to start using some woodchip as a main base for compost, possibly with other local materials such as bracken, seaweed and wool.

'We purchased a secondhand chipper a few years ago, but it's been very disappointing so this year we hope to buy a better chipper, sharing it with a few others locally who are also interested in using woodchip on their crofts. You can follow our progress online as The Wild Croft.'

woodchip to amend the soil health of the areas between the tree rows, you can even spread the ramial woodchip straight onto the ground as you go. You can either plant a single species in each row for a more even and efficient harvesting, or you can plant diverse species, which are likely to give better biodiversity. At Eastbrook Farm, we have interplanted our timber and fruit trees with willow and alder for coppicing. Though this is more complicated for managing the coppicing and chipping, our eventual aim is to end up with a line of large trees. The coppiced interplants give a temporary understory that make use of the space and give an output while the larger trees develop. We also hope that the nitrogen-fixing power of the alder will help the larger trees to grow.

Producing Woodchip at Garden Scale

For most home gardeners wanting a good source of woodchip, the simplest route is unquestionably to find a reliable external supplier. However, even small gardens produce woody material that needs to be dealt with. In my childhood, bonfires were the norm and I loved nothing more than wasting away autumn days staring into the flames. Though I have residual pyromaniacal tendencies, I do now recognise the environmental problems, and the sheer waste, of mass burning of carbon. In many places it is now illegal to have home bonfires. So, what are the best ways to manage this material on a small scale?

On a very small scale it is possible to manage it manually. You can cut tree branches with secateurs or loppers or lay them out on the ground and chop them up with a sharp spade or smash them up with sledgehammer. Invite your friends round for a chipping party – offer them a fish supper and call it fish and chips! This quickly becomes impractical, though, if you have even two or three trees or shrubs of any size.

In many urban areas in the UK, the local council will take your prunings away and process them into compost, which you can then buy back. I will cover the benefits and risks of using green compost in chapter 4. In most countries you have to pay for green material to be collected, whether that's through a government-sponsored scheme or by private contractors. If you have a lot of material this can get expensive and you are effectively exporting organic matter from your garden rather than investing it back into your system. In some less populated areas it may not even possible to get someone else to process your wood.

The next option is to use a chipper to process your own woodchip. We'll look in more detail at machinery options and the advantages of owning your own smaller machine versus renting larger equipment in chapter 3. It mostly comes down to the volume of material you are producing and space available to store that material pre-chipping. During the summer, most pruning, even of larger material, is fresher and greener. I tend to cut into 1-inch-long segments with secateurs straight into the compost bin. However, once I get to autumn and winter clearance, even in my small garden (20 metres by 5 metres / 65 feet by 16 feet) I need a powered solution. I have opted for a small electric shredder. I keep the larger branches to dry for kindling for the fire in the house and shred the smallest stuff for the garden. It doesn't give me very much usable material but does help manage it.

You might even consider clubbing together with neighbours to create a community composting system. The collaboration gives advantages of scale, prevents the wood being shipped further afield to be processed and turns the job into a social activity. In some areas there is support to help community composting projects. You might be able to work with a local tree surgeon, for instance, by providing somewhere for them to offload their chip, in return for them chipping your group's wood.

Free Woodchip

It is perhaps surprising that a lot of woodchip has little financial value, though the cost of production and transport are major contributors in this. There is a difference between carefully produced chip for biomass, where a specific product is manufactured for a market, and the chip that is generated as a by-product from cutting down trees for other reasons. Tree surgeons are paid to trim or cut down trees and need to do something with the wood that is created.

In the UK, a business that 'chips, cuts, shreds, pulverizes, burns, composts or stores waste plant matter' needs either an environmental permit or a waste management licence, or it can register as exempt from waste management licensing under some specific conditions.[12] If they:

- chip/cut/pulverize/shred less than 500 tonnes (1,000 tonnes in Scotland and Northern Ireland) in any seven-day period;
- store less than 500 tonnes (400 tonnes in Scotland) of waste material at any one time. There are some other regional variations – for instance, in Scotland farms processing only their own waste can get exemptions for up to 1,000 tonnes;
- burn less than 10 tonnes of waste material in any twenty-four-hour period at the site where the waste was generated, producing no dark smoke (Scotland and Northern Ireland only);
- spread biodegradable waste and soil from gardens and parks on agricultural land (conditions apply);
- use woodchip from untreated wood to surface paths.

In the US, though there is some variation between states, the picture is broadly similar with exemptions for small businesses. As you can see from this list, there is an incentive for small-scale operations to try and keep below the limits needed for licensing, or to make sure the wood is used on agricultural land or on paths.

There are larger wood recycling centres that will take arborists' waste, but they charge a gate fee. Arborists and landscapers may still choose to use these facilities; for instance, for larger loads or when there is wood other than just tree trimmings and logs, like fence panels or pallets. However, like all businesses they will be looking for ways to reduce cost. This is an advantage to those of us looking for free chip. Most tree surgeons are quite happy to give it away so they can get on with the next job and not have to store and manage it. Larger tree businesses have spotted the opportunity to add value, and if they have the facilities can create products from their chip that they then sell back to their customers for another income stream. This may become more common as the value of woodchip becomes more widely recognised.

There are some other specific situations where significant amounts of woodchip are generated that you might be able to take advantage of. For instance, power lines need to be kept clear of trees and the trees along the lines will be regularly felled or trimmed. Though some of this woodchip is left in situ, often this is not possible and the companies need to find a home for it. One energy cooperative in Indiana in the US, Hendricks Power Cooperative, offers woodchip to its members as a perk of buying the power from them. Other companies, such as Portland General Electric in Oregon in the US, have a specific policy of giving away the chip to the residents of the area they are working in and have a request form on their website to allow you to benefit.

Larger individual arborist businesses sometimes also generate enough chip for it to be worth working in a formal way with gardeners through their website to request chip. A quick internet search for arborists in your area may well provide a solution, but for smaller tree surgeons this is not always feasible. Consequently, websites have been developed to link up arborists with gardeners wanting chip, such as www.arbtalk.co.uk in the UK and www.getchipdrop.com in the US. The concept is very simple: if you want woodchip you register on the website, giving your address, and this is added to the list of potential woodchip drop sites. Arborists don't want to drive any further than they must, so will always be on the lookout for drop sites near their centre of operation. Even where companies have storage facilities or normal tip sites, they will often take on jobs outside of their normal area, and being able to dispose of the chip near where it was generated saves a detour and reduces load and fuel consumption on the journey home. Using these web platforms tree surgeons can check for their nearest drop site when they have a load to dispose of.

Usually, there is no charge for either party, though users can offer to pay for chip, which might push you up the list if you are keen to get lots of chip. There are options to choose what you are happy to accept and whether you need to be contacted first. If you have an easily accessible site, and it's handy for you to be there to supervise, or you are willing to allow tips unsupervised, you are more likely to get drops.

There are also community composting sites, often run by charities or the local government. Again, the chip from these sites is normally free if you turn up and load it yourself, or there may be a small charge if you need help getting it loaded up.

Things That Hitch a Ride with Free Woodchip

We have covered some of the potential risks from woodchip in chapter 1, but here it is worth just reiterating the point with regards to free chip.

Woodchip from tree surgeons with unwanted larger branches, Eastbrook Farm.

Of course, most arborists are decent, honourable people, and in theory should be looking for a long-term relationship with those receiving their chip, assuming they want to use your site again. However, even good operators have bad days or are tired at the end of a long shift. You don't want to lay yourself open to problems.

The larger operators clearing roadside trees or along power lines are likely to have the best-quality and most consistent chip. They will be clearing larger numbers of trees, often of a similar age and size. These trees also tend to be strong and healthy. They are being felled or pruned because they are in the way and growing well. Urban arborists and those working for private garden clients are more likely to be clearing dying or dead trees, trimming hedges and possibly clearing some garden rubbish at the same time. There is a greater risk of diseased material from garden contractors since this is often the reason the trees are being removed in the first place. This may not be a reason for not taking woodchip from them, but you might decide to treat it differently to a load that comes from a more reliable source.

The two tree surgeons that give us woodchip at Eastbrook Farm are generally great. Around 80 per cent of the loads are perfect, with just good woodchip. Some loads have larger branches or logs in them; this occasionally causes us some problems and we try to remind them what we need. However, on balance we value the chip enough and decide to deal with the problem. We have a fair amount of space and can put them to one side for chipping later. Just occasionally we get stuff accidentally in a load that becomes an issue; for instance, we have found a chainsaw helmet, a long pruning saw, a ladder and a bag of rubble.

I would recommend talking to the tree surgeon before they drop and ask them what is in the load. If your requirements are too strict, they may decide to drop at a different site, but that may be better than taking an unsuitable load.

Free Chipping Service

Many states in the US provide a free chipping service to residents where there is a fire risk from having trees and brash around buildings. Usually you will need to do the pruning and clearing first and prepare the material before arranging for the contractors to come to chip it. They will leave the woodchip at your property for your use.

Paying for Woodchip

What happens if you can't get hold of free woodchip or don't like what is on offer? It might not be available in the right quantity or at the right time, or the quality may not be good enough. You are at the mercy of the person giving it away to some degree. You might even be looking for something very specific, such as a single-species woodchip. What are the options for buying chip?

There are lots of landscaping suppliers that sell woodchip. Many of them have seen the potential for selling to gardeners and have a range of bag sizes, from carry-home 50 litre (1¾ cubic feet) bags up to 1-tonne dumpy bags. This is a pretty expensive way to buy chip, but it may be the most convenient if access is tricky or you live in a city garden and have nowhere to receive larger quantities, or if you don't have your own transport to collect free material. Prices vary widely: some quick research for the UK revealed a range from £30 ($42) to £170 ($240) for a 1-tonne bag. Some of these suppliers also offer truckloads, which obviously work out a lot cheaper than buying bagged. Assuming the companies are reputable the quality should be good and they may even have different grades available. As the customer you can expect a bit more. One company, for instance, even offers a crane service for moving chip over houses into inaccessible areas.

Tapping into the short rotation coppice, biomass supply chain can be a way of getting large quantities of relatively affordable chip. The trick is to find growers or suppliers that have off-spec material that can't be sold at biomass tariffs. Though prices vary, good-quality biomass currently sells off the farm for about £60 ($85) a tonne, which still compares well against the most expensive landscaping products but will likely be too expensive for most farming situations. However, there is material that doesn't make the grade. This might be the smaller chip that gets sieved out or loads that have been sitting around a bit too long and have started to compost. There are numerous examples of farmers that were persuaded to plant willow to supply power stations, which for whatever reason didn't work out. Many are looking for alternative markets and might be willing to sell for less, especially if you can arrange for the chipping and collection of the material. It may take a few phone calls and a drive about some farms, but the result could be a mutually beneficial arrangement.

Woodchip for Animal Bedding

While straw is the most used material for animal bedding, woodchip can provide a very successful alternative. Straw is the by-product of arable crop production. In traditional mixed farming systems, it makes sense that the straw should be used for bedding and then returned to the land incorporated with the animals' excrement to enhance soil health and fertility. In more recent times, farms have become more specialised, with the majority of larger arable operations taking livestock out of their rotations. In these cases, the straw should arguably be left in the soil to maintain the soil structure and fertility, but with margins low there is a constant pressure on farmers to make money from all potential outputs. As a result, much of the straw is sold off-farm to specialist livestock farmers. Sometimes the cropping farm will get the manure back, but often it doesn't, and they will rely on bagged potassium and phosphates and the organic matter of the soil gradually diminishes until, in worst-case scenarios, the land becomes unusable for growing crops. Using woodchip instead of straw allows us to use a material from more marginal, less productive land to sustain soils on more productive land. Not all farms, smallholdings or backyard chickens are in areas that grow arable crops successfully. This can make getting enough straw difficult, but even where it is available, using woodchip can save on the energy and financial cost of shipping straw across the country.

Using woodchip has some advantages over straw, but there are a few things to look out for. Since this is not a book on animal husbandry, I don't intend to cover all details, but I will briefly outline some of the potential of using a woodchip-based bedding system. Much of the information here comes from the Woodchip for Livestock Bedding Project run in Wales.[13]

There are a few things to beware of when using woodchip for bedding:

- Only use untreated wood; anything with paint or preservatives can be a risk to animal health, as can the dust from products like MDF.
- Avoid old pallets or similar, as bits of metal or nails left in the material could also be harmful to animals.
- There are some species of wood that are best avoided, too; for instance, black walnut can be toxic, though if left for a month exposed to air this risk reduces. Larch is not recommended as it can splinter, making it sharp for the animals to stand on. Similarly, thorny species can cause injury, particularly blackthorn.
- Woodchip from landscapers and tree surgeons is also often best avoided, since it can contain a proportion of poisonous species such as walnut, rhododendron and yew.

Otherwise chip from most tree species appear to work well, with some species particularly recommended. For instance, spruce is reputed to be soft with plant compounds that suppress harmful fungi and bacteria. The type of chipper used, as well as the moisture content, can have an impact on how much the wood splinters. Robert Myers Paul found that woodchip with splinters formed a more compact bedding

Sheep with woodchip bedding in Scotland. Photo courtesy of Dr Audrey Litterick.

layer that drained less well.[14] Larger particles with a blockier square structure filtered better, allowing the top few centimetres to remain dry, leading to improved animal comfort and welfare. If the quality of the chip is good enough, though, animal welfare should be as high in woodchip systems as in more traditional straw bedding.

Apply the bedding in shallow layers of about 10 centimetres (4 inches), topping up with a dry layer when needed, usually every seven to ten days if the animals are being fed on a dry diet. Older wood is more suitable for bedding as it tends to be drier and higher in carbon; this means it will absorb more moisture from the manure and hold onto more nitrogen than ramial chipped wood. Most woodchip has a similar water-holding capacity to straw, though it is crucial to keep the moisture content of the chip below 30 per cent or it will lose that absorbent quality. The need to keep the material dry has implications for how you manage it. For example, storing it unchipped takes up less space and stops it absorbing so much moisture.

A European study comparing woodchip and straw bedding found that, 'the wood chip treatment resulted in somewhat cleaner animals,'

Case study

Peter Clifford is a tenant farmer at Hey End Farm on the Dartmouth Slaithwaite Estate in Yorkshire in the UK. He is seeing immediate benefits to his soil from using woodchip bedding compost, as well as experimenting using a natural way to cover the heap and reduce leaching.

'I have been successfully using woodchip as a winter bedding material for my cattle. I've found that if then stored in a compost heap for a further year, having been mixed with my cattle's manure in the shed, it forms a fantastic rich compost. I initially seeded the surface of the compost pile with new grass seed, which grew vigorously all over the pile and, when I subsequently spread the compost in the field in the following year, I added further new grass seed direct to the muck spreader. After spreading, the manure was rolled down to allow the new seeds to take and germinate.

'I first applied this compost to two fields that were originally very poor and reverting to moorland – that is, bilberries, heather and moor grass. Shortly after application the fields burst into green life and are now producing very good grazing. I did not need to apply any chemical fertiliser. Poor productivity of the land limits the number of animals that can be sustained, meaning that producing sufficient manure from winter housing alone to improve the land is difficult. However, the composted woodchip provides much additional bulk and nutrients. I'm sure things will only get better year on year and so I am keen to continue use of woodchip in this way.'

but that, 'the heifers had a significantly longer total lying time when kept on straw.'[15] Looking at some of the farming forums on the topic suggests that while there is significant potential to replace or supplement straw with woodchip, it is not always straightforward getting the system to work. Full-cycle financial analysis is also worth looking into. While straw might often be cheaper, its prices tend to fluctuate according to availability while woodchip prices remain relatively stable, particularly if you can produce your own.

Using Woodchip from Livestock Bedding Systems

The Woodchip for Livestock Bedding Project considered the potential for using the composted material after it had finished life as bedding. Not only is this desirable if aiming for a low-input system, but stacking benefits from resources makes good business sense. One possible option is to reuse it as bedding. It can be used two or three times; each time it absorbs more nitrogen. More research is needed into the differences in performance between various tree species.

The composted chip can be used in the same way as a traditional straw-based manure, and using the same machinery, by spreading on fields to improve fertility and soil health. Due to the smaller particle size it can be easier to handle and spread than straw-based manure. One Canadian study found that woodchip bedding lost less nitrogen and carbon during composting, while another found that composted woodchip manures leached less nitrogen (and some other chemicals) than barley straw manure once added to soil.[16] This suggests that woodchip may offer potential for better resource use, longer nutrient retention and reduced risk of water contamination.

One word of warning though: because the chip takes longer to break down than straw, bits of uncomposted wood will remain in the material, which can cause problems in slurry systems, for instance in pumps from slurry pits and in the slurry tankers. Woodchip bedding may, therefore, work best in solid manure management systems.

•

CHAPTER THREE

Managing Woodchip

Gearing your system up to manage woodchip depends a lot on how much you need to process. At one end of the scale, you might just be looking at bringing in the occasional bag for home use, where storage and management are minimal. Farmers with their own trees to process may need to create dedicated areas where they can properly control the composting process and material.

Chipping

Let's look first at the actual chipping process. There is a distinction between chipping ramial chipped wood (see page 109) for immediate use and converting larger-diameter wood. The former will need to be chipped or shredded and spread quickly, while the chip from older wood is normally stacked and composted for a while, unless you intend to grow mushrooms on it.

There are some basic questions to ask yourself:

1. How much wood do you have to chip?
2. How big are the trees?
3. How much space is available to stack the wood pre- and post-chipping?
4. How easy is it to either get the trees out or the chipper in to where you need to work?

If the answers to all these questions are 'lots, big, lots, very', I'd recommend hiring a massive chipper once a year to do it all in one go. If, on the other hand, the answers are 'not many, small, hardly any, not at all', you might decide to buy a smaller machine that is on hand whenever you need it, or regularly hire in a smaller machine. There will be other factors, such as whether you can find a decent secondhand chipper to buy or how easy it is to find a contractor to chip for you. Access to the site may also be a challenge if you don't have good roads or big gateways.

As a rule, if you have the space, I recommend piling up your wood to be chipped and hiring in a big chipper to do the job in one go, rather than buying a chipper. You will normally be able to afford to hire a much bigger chipper than you can buy, and the job will be quicker and work out cheaper in the long run.

Leaving wood piles around can pose a fire risk, especially in drier climates. Make sure that any piles left for chipping are stored well away from buildings and machinery. Some areas even offer a chipping service to residents to reduce the risk in urban areas of piles of garden debris. For instance, the Park City Fire District in Utah offers a free pickup service in the summer months.

For very big operations, or on estates that have significant chipping requirements, it may be more cost-effective to buy the machinery and use it to take on contracting chipping for other farms to help offset that cost.

When to Chip

In terms of the quality and usefulness of the chip, you can cut wood at any time for most purposes. There are exceptions, such as using willow as a disease-preventative mulch on trees, where timing is critical to allow for the maximum release of the salicylic acid. Similarly, if you are chipping for mushroom production, or using ramial woodchip for soil health, you need to have the material freshly chipped in spring or fall to use immediately. For most other uses, you can cut when it is convenient for you and either chip immediately or pile up ready for chipping at a future date. The wood will dry and season (or slowly compost) but that is not usually an issue.

Receiving, Stacking, Turning, Storing

Whether you are chipping your own wood or receiving loads from elsewhere, you need somewhere to put it and a system for processing it. Though it may appear to be more stable than, say, animal manures, which have a visible brown runoff, woodchip does also release leachates

Pollarded willow material stacked and ready for chipping. An ideal habitat for small children.

when left out in the rain.[1] This is much more likely to be a problem in large commercial yards or where significant-sized piles of woodchip are left after processing. However, if you are planning to use or process woodchip on any scale, be mindful of where that runoff is going. The leachate has been shown to be harmful to aquatic life, so do not site woodchip piles near water, including garden ponds. This will not be a problem for using the woodchip as a mulch, but if you had a large sack waiting for use in the garden, I wouldn't store that next to my pond. If you want to produce a holistic system you could grow some cop-piced willow around your chip site, as a bioremediate for any leachate, chipping the resultant coppiced material to go back into your system.[2] My experience is that in most conditions the chip will absorb most rainwater and it is only in very heavy precipitation that runoff occurs.

Infrastructure and machinery depend hugely on scale. At the small end, where you simply dump the woodchip in a pile and move by wheel-barrow or small trailer, so long as you are not directly next to water, putting the chip onto grass or a concrete pad is fine. If you need to subsequently move the chip with a front loader, you will need either a well-drained firm piece of ground or, even better, a hard standing area.

Chipping yard showing unchipped and chipped wood.

If you don't already have this type of hard area on the farm, it can be a big investment with a hefty environmental impact: around 150 kilograms (330 pounds) of carbon are estimated to be emitted for every cubic metre of concrete produced. For both cost and climate benefits, if we can find a way of managing chip without concrete, so much the better. A small digger is often a better turning option for medium-scale turning, offering more flexibility and reduced pressure on the soil.

As we explore in other sections, woodchip will gradually break down as the fungi work their magic. We can speed this process up by turning or adding some higher-nitrogen material, but this takes time and energy, and at scale requires turning machinery and a suitable site. An alternative is to identify where you are hoping to use the chip and drop piles of woodchip there. If you need them to be well rotted, you will need to leave them in situ decomposing longer than you would have to if you were turning first and then applying. This still may be an easier option for those without machinery or on very heavy ground.

It's worth investing a bit of time in strategic planning on how and when to store and move chip around the site. Winter is usually the best time for chipping, partly because that's when you are likely to be

Woodchip piles ready for mulching around a new tree planting. Leaving chip near the place of use reduces wasted travel between loads.

cutting trees, but also because other jobs on the farm tend to be less pressured. However, winter is often also the time when fields are wet and moving heavy loads about is not a great idea. If you get long periods of freezing conditions, of course, this is not a problem, but in the UK we are getting fewer hard frosts and rarely for long enough to freeze the ground solid. If storage space is a problem, summer chipping when the ground is hard may be more suitable, moving the chip to the site immediately where it will be used, even if it sits there for a few months before spreading. When chipping your own trees, it may be best to leave the chip in piles where you chip them, let them rot down for a while in situ and come back to them when you are ready to spread. This will save double or triple handling, all of which adds to cost.

If you get large amounts of chip from tree surgeons, you will need a fairly good receiving area, or they won't be able to dump them. Depending on how often you get deliveries, you'll need to keep on top of managing the pile. At Eastbrook Farm we get at least two loads per week. Our concrete receiving area is not huge so if we don't get the loader to consolidate the pile regularly, we quickly run out of space. As we move the piles, we also take the opportunity to turn the rest of the

Freshly delivered woodchip before turning at Eastbrook Farm.

pile. We have gradually accumulated a mound of logs and branches. These occasionally need to be chipped, though we have also been using some of them to create log piles for habitat around the edges of our agroforestry fields.

How Often to Turn the Heap?

This depends on:

FINAL USE

If being used for mushrooms, you want the chip as fresh as possible and unturned. For use as a mulch on mature trees you can also apply unturned and fresh chip, though it is also good rotted down. For no-dig vegetable systems, mulching on small plants or spreading for soil health, turning will speed up the fungal decomposition and ensure that all of the material is composted. You can also just leave the heap for longer if you have the space and the result will be similar. You might have some larger pieces left on the top of the heap and at the edges that have not broken down fully, but these cause no problems once incorporated with the rest of the heap.

COMPOSITION OF THE CHIP

Larger chip from older wood will need more turning and longer to break down than small chip or ramial woodchip. If the heap contains

Loadall turning woodchip showing steam from hot composting.

any sawdust or very small particles there is a risk of creating pockets of the pile so compacted that no air can get in. This will stop the composting process, so turn more frequently. Similarly, if there is green material mixed in with the chip, or you add manure or other high-nitrogen materials that will also speed up the composting process, you might need more turning to ensure a good distribution of the different ingredients.

HOW QUICKLY YOU NEED THE CHIP TO BE READY
Turning speeds up the composting process. If you need the space or the material more quickly, you can spend more time turning. If you can wait, you might decide to let nature take its course or just do one turn to make sure the material is well mixed.

Composting Woodchip

There are broadly three ways to compost woodchip. You can simply leave it in a pile and wait for nature to do its thing. If you have no machinery to aid you, and time and space are not pressured, this approach will still give worthwhile material. In most cases, though, a more proactive method is preferable. To actively manage our woodchip piles, we can choose between a traditional turning technique and a static 'forced air' arrangement. Before we get into the details of different composting systems, however, let's go back to basics and see what is happening in our woodchip pile, to see how best to harness those biological processes to make the best material.

In a traditional composting system with garden material we aim for a good mix of so-called green (high-nitrogen matter like vegetables or leafy material) and brown (high-carbon woody debris or cardboard) material. Since even green matter has quite a lot of carbon in it, in practical terms this usually means mixing two parts green to one part brown, looking for the magical carbon to nitrogen ratio of 30:1. The trick, then, is to control the temperature of the heap by turning to prevent it getting too hot, which would kill off all the beneficial microbes, but ensuring all parts of the heap reach a high enough temperature (60–70 degrees C / 140–160 degrees F) to kill weed seeds and pathogens.

Woodchip composting works slightly differently. Firstly, by its very nature woodchip is much higher in carbon than a garden compost; typical 600:1 rather than 30:1. Secondly, wood has a structure that bacteria find almost impossible to enter and degrade. To succeed with woodchip we are relying instead on fungi, which not only have the ability to penetrate the wood's tough structure but can also function with lower levels of nitrogen.

Most woodchip should break down in three to twelve months. In the UK, we generally get enough rain that we don't need to water our heaps. The composting tends to slow right down during the hottest summer months but starts up again as the temperature drops. In hotter, drier conditions you will likely need to add moisture at certain times to keep the fungi active and the composting process happening. Water tends to run quickly through the heap, so a very gentle sprinkler system that allows the water to soak into the material is best. You may also choose to cover the heap to conserve that moisture and shade it from the hot sun.

On page 62 we'll look at the Johnson-Su composting bioreactor as a specific method of composting woodchip that can not only overcome these environmental factors but also produce a more biologically diverse product than a windrow-turned method (see page 61).

Keeping It Low?

There is some debate about the best size for a pile or windrow of woodchip. Some recommend 1 to 1.5 metres (3 to 5 feet) high, while I have also heard reports of very successful 3 metre (10 feet) high heaps. To find the right size for your system, consider what space and machinery you are going to use, but also think about your climate and how you will create a good environment for the fungi. It needs to be large enough for the centre to remain cool and shaded. Larger piles run the risk of getting too hot in the centre and self-combusting; even if the pile is safely away from buildings and machinery, a burnt pile of chip is not much use. In damper conditions, however, this is unlikely to happen.

Make a cup or dip in the top of the pile or windrow to help any water filter through the middle of the heap rather than run off down the sides. This isn't usually a problem when the chip is very fresh and there are lots of gaps, but as the fungi starts to break it down it will become more compact and the surface for the heap can form a barrier that prevents water infiltrating.

Screening and Sieving

Depending on how we plan to use the chip, and on how big and uniform it was to start with, we may need to do a final processing stage. You won't need to screen if the chip is destined for rough mulching

Small-scale sieving and turning equipment at Tolhurst Organic on the Hardwick Estate in Berkshire, England.

and you are spreading by hand. However, some of the chip we get in still has a few large pieces of wood in, which if put through the mechanical spreader would block it. Some research has been done in commercial biofuel operations to develop screening tools and look at the cost benefit of increasing the value of the chip.[3] I have also seen one operator in the UK put a homemade solid wire mesh screen over the hopper of his side-discharge spreader to catch the larger particles as he loaded the material.

When used as fine mulch or for potting and propagation compost you will definitely need to put it through a sieve. On a small scale a garden sieve or plastic crate can work well; on a larger scale I've had success with a rigid wire mesh over a trailer. Alternatively, mechanical options are available. For instance, the German company Scheppach produces a range of rotary soil sieves that are well worth the investment if you are planning to make a substantial quantity of growing media. Larger operations will of course use more substantial versions of this type of machinery. Final use determines the size of the screen. The finer the mesh, the slower it is to get the material through, so choose the biggest mesh size you can get away with to achieve the texture you need.

Propagation woodchip compost covered for final maturing phase. Covering keeps the weeds out but can create anaerobic conditions.

Storing

Woodchip, whether composted or not, is a fairly stable material and will happily sit without coming to much harm until you are ready to use it. The longer you leave it, of course, the more it will decompose, and you will be missing out on some of the potential benefit. Ideally, we want to use it as soon as it reaches the desired state for that purpose.

There are other risks to leaving piles lying around too long. Notably the risk of weeds. Composted woodchip is an ideal medium for germinating seeds, and it doesn't take long for weed seeds to land on the pile and start growing. As long as the weeds don't germinate and you are intending to use the chip as a mulch or for spreading on fields this might not matter, but in practice the difference between a cover of smallish weeds and 'whoops, they seem to have seeded all over my heap' can be a matter of days. Groundsel (*Senecio vulgaris*), for instance, can complete its entire life cycle in five weeks, and even if you pull it up in flower it can continue to form seed as it dies.

For heaps that you are not ready to use but no longer want to actively manage, the best thing is to cover them; if you have spare barn or shed space you can move the composted chip indoors. Alternatively, you can cover the heap with a tarpaulin or thick plastic to keep it from

Wildlife making their home in a woodchip pile.

weeds and leaching, though there are issues with this method as it cuts out air from the heap, which will have an effect on the beneficial microorganisms in the pile. Breathable covers or no covers at all are better than plastic for the health of the compost. If you do want to use plastic to reduce weeds, then cover only the top of the pile. You can buy special windrow covers made of a breathable fabric if you intend to invest more heavily in your system. Properly covered, the composted woodchip can last well for a few years, though it does provide a lovely home for mice, rats and other little creatures.

How Fungi Break Down Wood

Wood does not rot easily; this is because the enzymes that normally break down the natural polymers need nitrogen to be able to form. Lignin is the toughest of these and makes up about 30 per cent of most woods. It forms a barrier around the cells, making it harder for the enzymes to reach the cellulose and hemicellulose polymers that are easier to degrade. Woody plants also produce their own defence against fungi; for deciduous trees these are normally tannins, while coniferous species manufacture thujaplicins and a host of other volatile and complex compounds.

There are, however, three types of fungi that can break down wood:

1. **Soft rot fungi** have evolved in damp conditions to make use of the low levels of nitrogen within wood, or, if that is not available, they can scavenge it from the environment, usually from the soil. This is one of the reasons that fence posts rot in the part that is in the ground.
2. **Brown rot fungi**, by contrast, use an oxidative process to produce hydrogen peroxide, which breaks down the cellulose and hemi-cellulose. The lignin is left pretty much untouched, which gives a hollow brown structure to the wood, hence the name.
3. **White rot fungi** are the only organisms known to be able to process lignin. They make super-efficient use of the limited nitrogen in the wood to produce enzymes, which in turn manufacture very strong oxidants capable of degrading the lignin.

So, we can see that availability of nitrogen (as well as phosphorus) is a major influence in the decay of wood. The two factors we can use to over-come this are time in the heap (leaving the woodchip longer before using) and the age of wood we use. Younger wood, in smaller branches, bark and sapwood, has higher levels of nitrogen in it, allowing for quicker rotting. Heartwood has the least; adding a source of higher-nitrogen material, like manure, helps to kick-start the process. If you are planning to mix manure as part of the composting, the general recommendation is about one part manure to three parts woodchip, though of course this will depend on the nitrogen content of the manure and the nature of the wood.

Types of Composting Systems for Woodchip

The simplest way to actively compost woodchip is to pile it up and turn it regularly. At garden scale you can do this easily by hand, either moving the heap from one spot to another, or if space is limited simply moving the pile about so that any material that was on the edge is moved to the middle and vice versa. Larger piles can be turned with a front loader or digger as described later in the machinery sections on page 65. For those with a scientific bent, or if you need to get the materials composted quickly and moved on, monitoring the temperature and turning at critical points helps to ensure the woodchip is fully composted in the shortest period. For most purposes, fitting in the turning around your other priorities works fine. Three or four times a year should suffice.

Windrow Turning

Creating a windrow is one way to scale up your composting pile and makes better use of space than several individual piles. It is a method which creates a long pile, rather than a big heap, of material, usually 1.5 to 2 metres (5 to 6 feet) tall and around 4.5 metres (15 feet) at the base, though they can be bigger. There is specialist windrow turning equipment available, which operates as a giant Archimedean screw, effectively moving all the woodchip around with minimum effort. It ensures that no material is always at the outside or inside of the heap. If you are planning to go into woodchip production on a big scale, it is worth considering a windrow turner, but for most of us the machinery on farm will do a good enough job. If you have the space, the windrow system is still a good one; the pile will of course get smaller as it composts, so the idea is to consolidate the material but keep the height and width the same, while your row gets shorter.

Aerated Static Composting

You don't need to turn the woodchip to have a good composting system, however. There are several options, sometimes termed static methods, for composting that work well for woodchip, particularly if the material is fairly consistent. If a pile of mixed woodchip is just left it is likely there will be some areas with less nitrogen or some more compacted patches, which in a turned system would be broken up and moved about. The key in a static system is to ensure there is a good supply of oxygen and moisture to allow fungi to colonise the whole heap and facilitate the subsequent migration of bacteria and other microorganisms into the decaying wood. This can be done through either passive or active aeration. The former creates channels within the heap to allow air to move through, while the latter uses a blower of some sort to force air through the rotting woodchip.

The most well-known passive aeration system is the Johnson-Su Bioreactor. This controlled composting system is named after David Johnson and Hui-Chun Su. There are some very clear instructional videos available online demonstrating how to create the heap, but in simple terms their method involves inserting a series of plastic tubes through a carefully constructed heap placed on a pallet lined with landscaping fabric. The whole heap is encased in a cylinder of wire mesh, again lined with landscape fabric. The construction and filling of this bioreactor, though not complicated in engineering terms, will take longer than just dumping up your material. The woodchip has to be pre-wetted, and loading the chip around the tubes requires some care. However, once built and filled your job is pretty much done until the composting has finished and you empty it. In the New Mexico climate where they developed their method, irrigating is key to success, and so they include an automatic watering system in the design. In damper, cooler climates it should be possible to irrigate far less frequently, and by hand, and still get good results.

The main purpose of using a bioreactor versus simpler composting methods is to create a biologically superior compost. Johnson and Su found that because the fungi were not being disturbed by turning, their output had higher fungal levels than turned compost. The diversity of fungi was also greater than that in a turned compost. We will explore in more detail in chapter 5 how we might be able to exploit that fungal richness to boost our soil health.

Though less relevant when using pure woodchip as a feed source, Johnson and Su also found that when composting dairy manure, where high salinity levels are a challenge, their composting process resulted in a finished product that was lower in saline. If you are planning on mixing manure with the woodchip, this would be an added incentive to look at using their methods.

While many composting methods will boast about how quickly you can produce a finished product, Johnson and Su believe a slow process gives a better result. Their tests showed that their compost at twenty-two weeks had 424 species of microbes, but at sixty weeks that had increased to 453 species present. Their recommendation is to take the process to at least fourteen months, though it is likely there will be variation according to climate and material used.

Active Aeration

The method of blowing air through compost to speed up and improve the composting process originated in the 1970s with the US Department of Agriculture. Unlike the Johnson-Su approach, which aims never to

Ben Taylor-Davies has been developing a farm trailer bioreactor design. Photo courtesy RegenBen at Townsend Farm in Herefordshire, England.

move the composting material, in most active aeration methods the aeration is used only in the first part of the process to speed up the initial breakdown. Often the material is then turned and left for a more prolonged maturation phase.

A typical active aerated system includes a loose woodchip bottom layer on top of the aerated pipe to ensure that the air can get into the rest of the heap. The main material is then piled on top, either in a typical pyramid-shaped stack, or if doing 'in vessel' just filling the container up. This is topped off with a 'cover layer' of a finished compost. This top layer helps insulate the heap, reduces odours and keeps moisture in the pile. It is also typically full of a wide diversity of microorganisms, which will colonise the compost as it cools and matures. There is no reason that we cannot adapt these methods for woodchip composting.

As we have seen with passive systems, moisture content is key. In turned systems, wet and dry layers or patches are regularly mixed, but in static piles we don't get that opportunity, so we need to make sure the material starts at about 50 per cent moisture content.[4] We also need to ensure the pile doesn't dry out. Once it has, it is almost impossible to rewet without turning. Even in damp English conditions, in big heaps I have come across bone-dry patches at the bottom. Rainwater or irrigation from the top just doesn't get to the base of the heap.

You won't need to pump air constantly into the stack, and indeed that will lead to problems, most notably potential drying out. In addition, actively controlling the aeration in relation to the temperature of the heap gives a better final material. One study found that the fungal population was substantially larger in the compost that had been actively managed compared to one that had received constant aeration.[5]

Harnessing Other Outputs from Decomposing Woodchip

In chapter 4, I explore how to use the warmth from a rotting woodchip pile for propagation, but on a larger scale there is potential to capture and use efficiently the gasses and heat that are emitted during the decaying process.

Jean Pain was an engineer who pioneered harnessing the outputs from woodchip decomposition half a century ago. Together with his wife, Ida, he spent many years developing a beautifully low-tech system that took the readily available brushwood around their farm, and as it decayed saved that energy to heat the water for his house by passing pipes through the rotting woodchip, and in the process collected the methane released to use for cooking. Ida Pain wrote the wonderful book *Another Kind of Garden* (still available as an e-book and highly recommended) outlining Jean's methods and their philosophy. The many pictures of their endeavours show their pioneering attitude, as well as a scant regard for health and safety – for instance, the image of Jean scything in nothing more than a pair of swimming trunks. However, this is also a practical guide with detailed facts and figures on what they achieved. In this book, they record that their experiments 'have shown that a 50-ton heap is capable of producing hot water at 60 degrees Celsius [140 degrees Fahrenheit] – it entered at 10 degrees Celsius [50 degrees Fahrenheit] – at a rate of 4 litres (1 gallon) per minute, without this interfering with or harming the compost'.

He created a larger heap: 5.5 metres (18 feet) in diameter and 3 metres (9 feet) tall, and by packing the material tightly slowed the composting rate so that he was getting heat and methane for 18 months before finally dismantling the pile and using the composted chip in his vegetable garden. He was even powering his old Citroen from the gas produced.

This process is not true composting, which should be a totally aerobic process and produce no methane. Methane is a product of anaerobic respiration. I would advise against any open woodchip composting system which produced methane even for part of the process, given that methane is a damaging greenhouse gas. This is different from an anaerobic digestion system where the methane is captured and used. There is also a serious health warning with any low-tech collection of methane since it is highly flammable and there are regular reports of people blowing themselves up while generating their own methane for cooking.

Machinery

Woodchip is a bulky material, and one that requires some force in turning a tree into small bits of wood. Even at garden scale, some machinery is usually needed to produce woodchip, and as the quantity of woodchip used increases, so does the need for equipment.

Garden Scale

I will cover small-scale woodchip production and use separately since most of the larger machinery will be of no relevance. However, there is no specific size beyond which you suddenly need to invest in a particular bit of kit. Some of the medium-sized chippers are a great purchase if you can afford them, even if they do not get fully utilised for your scale of production. Similarly, at Eastbrook Farm we have spread a lot of chip with barrows, as it offers a safe job for volunteers where we can't have them all jumping untrained onto the farm machinery.

CUTTING

Most gardeners will choose secateurs, loppers and pruning saws as cutting tools. For bigger trees, chainsaws are necessary, and for larger quantities of wood they are a good way to speed up work. However, unless you have a chipper that will take the larger wood, you might just be building up a store of unchippable wood – great if you have a fire with an appetite, but otherwise a material that needs disposing of. You might decide it is easier to bring in a tree surgeon once a year to cut and chip everything.

CHIPPING AND SHREDDING

Firstly, what is the difference between a shredder and a chipper? A shredder works with the grain of the wood, rather like an axe does when splitting logs. It splits and shreds smaller branches and greener wood. The blades or flails are semi-sharp and depend on generally mashing and smashing the wood up rather than cutting it. Shredders are less prone to clogging than chippers but are more limited in what size of wood they can process. Chippers, on the other hand, work with sharper blades, which cut wood up into chips. They can deal with bigger and harder wood than shredders, but often get jammed up if there is too much soft green material put through them.

You can pick up an electric garden shredder secondhand for £20 (under $30), but understand the limitations of this type of machine. It will shred very small brash – usually a maximum 1.5 centimetres (½ inch) – up into a usable material. It is slow and tends to produce a long, stringy product. I do have one in my backyard and use it three or four times a year just to clear my light prunings. It doesn't produce much chip, however, and is more of a tool for keeping the garden tidy.

If you are serious about producing woodchip, you will need to go up a level to a proper petrol- or diesel-powered chipper. There are already some more powerful electric models being developed and no doubt these will eventually replace the fossil fuel models. There is an increasing range of domestic-scale products to choose from, starting at around £400 ($550). They are standalone, so you can take them anywhere in your garden rather than being reliant on an electrical supply. They'll also deal with bigger material; the cheaper models claim they can chip 4.5 centimetre (about 2 inch) diameter material, and as the power and size of the machine increase so does the size of branch it will chip. For around £1,500 ($2,100) you can get something that might dispose of 8 centimetre (about 3 inch) diameter branches. Whether to buy one of these types of machine or to hire in a chipper occasionally will depend on how much material you have, and what annual use you have for larger branches. If you can use the larger logs for a home fire, being able to chip the cuttings is sensible. Hiring a chipper able to cope with much larger material might be more cost- and time-effective than buying smaller equipment.

HARVESTING AND MOVING

At garden scale, you are most likely going to just chip directly into a wheelbarrow or pile in a corner to be moved later. Obviously, for larger gardens a trailer with tall sides, to hold more, makes moving chip more efficient if you have something to pull it.

TURNING AND SPREADING

For smaller heaps, turning is less crucial. Use a fork to move the material around or a shovel to move the heap. For spreading, a barrow or small trailer is the most effective way of moving chip. It is a light material, so building a frame for your barrow to allow you to pile the woodchip higher will save time in spreading if you have a lot of material.

It's worth just mentioning which hand tools are most effective for moving woodchip. Most of us have a shovel or two, and these work fine, but they tend to be quite heavy. There are lighter versions; scoop shovels or snow shovels, for instance, which work well. They allow you to get more woodchip with every load, and since woodchip is very light

Paul moving woodchip with a manure fork at Eastbrook Farm.

it shouldn't kill or incapacitate you. However, all these shovels only really work if you have a hard surface and you scrape along the ground under the heap. Using a bedding fork means you can get in at any height, which saves so much bending over and the inevitable resulting back pain. When we replaced the shovel with a fork at Eastbrook, my colleague and good friend Paul Clark had a new spring in his step.

Farm Scale

The principles of creating woodchip remain as scale increases, but you have the option, or in many cases the necessity, to bring in larger machinery to process and move the quantity of wood generated. Some of that kit may already be available on-farm, like front loaders and trailers, but you will likely need to buy or hire much of the specialist machinery.

CUTTING

On a small commercial scale, a chainsaw is still the most versatile tool. It will give the neatest cut, can be taken to the most inaccessible hedges and work around those that are full of wires and old fences. For hedges less than 100 metres (330 feet) long or small numbers of trees, the chainsaw is also likely to be the most economical option. The cost of hiring larger equipment won't be justified by the time saved. I know three people who have lost parts of their body to chainsaws, and working in tangled hedges is particularly tricky. Getting proper training and always wearing suitable protective gear is essential with chainsaw work.

For hedges more than 100 metres (330 feet) in length or large numbers of trees, these are some of the other options for cutting.

Tractor-mounted circular saw. Crude but effective, this will make short coppicing work of most hedges. The drawback is that you can't control where the material falls; this makes it less suitable for bigger trees and it can also make it difficult to operate on its own. Having a second tractor or telehandler with a grab and working in tandem moving material once it's cut is more efficient.

Tree shears act like giant secateurs; this makes them easier to manipulate and more accurate than the saw. Since they also usually have an integrated grab to remove the material to one side or straight into the chipper, they can also act independently. On the minus side, they tend to be mounted on excavators, which makes them big and heavy with associated soil damage and access limitations.

Specialist felling heads for excavators are also available; these have integrated circular saws or chainsaw cutting bars, with grabs or accumulator arms.

As part of their extensive hedge management machinery trials, the Organic Research Centre collated some quantitative results shown in Table 3.1.[6]

There is also an excellent report focusing on wood fuel extraction from hedges in Devon in England, with some detailed costings from individual farms with hedges in different states of growth. This report notes the efficiencies of a more mechanised approach: 'It is far quicker to process the wood if all the material taken from the hedge is chipped on site. 100 m of hedge can typically be chipped in 1–2 man-hours. In contrast, it typically takes 3–4 man-days to extract the cordwood, cut and split it ready for the stove or boiler, and burn the brash on site in a bonfire.'[7] Though the report is aimed at biomass production for fuel, even if you don't plan to burn the wood, much of the information in this report is useful for informing management decisions.

Table 3.1. Comparison of Costings for Coppicing Hedges Using a Range of Different Machinery

Machinery Description	Option Trialled
Hydraulic shears	Excavator mounted Dymax 250 millimetre (10 inch) grapple tree shears with added accumulator or feller buncher functionality mounted on an 8 tonne Komatsu PC78MR-6 zero swing excavator; 1 man and 1 machine
Hydraulic shears + chainsaw finish	Excavator mounted Dymax 250 millimetre (10 inch) grapple tree shears with added accumulator or feller buncher functionality mounted on an 8 tonne Komatsu PC78MR-6 zero swing excavator; 1 man, 1 machine and 1 chainsaw
Timber grab with integral chainsaw	Excavator mounted Gierkink GMT 035 felling grapple with chainsaw cutting bar mounted on 5 tonne Kubota excavator; 1 man and 1 machine
Single circular saw – hedge 2 Wakelyns	Tractor-mounted single circular saw attachment on hedge cutting arm, with second tractor and front-mounted fork; 2 men, 2 machines and 1 chainsaw
Single circular saw – hedge 1 Wakelyns	Tractor-mounted single circular saw attachment on hedge cutting arm, with second tractor and front-mounted fork; 2 men, 2 machines and 1 chainsaw
Manual fell – hedge 1 Wakelyns	Motor manual extraction of cordwood for firewood logs: man with chainsaw to separate brash from logs & prepare logs for firewood; brash to be chipped. Tractor with front-mounted fork used to separate pile of hedgerow material
Assisted fell technique	Motor manual felling supported by an 8 tonne Doosan DX80R excavator with a front-mounted 1.8 metre (6 foot) land rake and Husqvarna 390 XP chainsaw with 600 millimetre (24 inch) cutting bar; 2 men, 1 machine and 1 chainsaw
Manual fell – hedge 21 Elm Farm	Husqvarna 560 XP chainsaw with 425 millimetre (15 inch) cutting bar; 2 men and 1 chainsaw

Minutes per Metre	Metres per Hour	Metres per Day	Day Hire Cost (£)	Haulage Cost (£)
2.78	21.58	151.08	525	500
3.3	18.16	127.12	525	500
2.64	22.73	159.11	500	500
2.9	20.69	144.83	480	100
5.4	11.11	77.88	480	100
12.85	4.67	32.68	224	0
1.58	37.97	265.82	450	150
10.81	5.55	38.86	300	20

CHIPPING AND SHREDDING

Once you have your material, what are the best options for chipping? For garden use, a shredder is usually more suitable, while for farms that are likely to have more mature trees, chippers will be the better option. (For an explanation on the difference between shredders and chippers, see page 66.) For most uses, shredded and chipped timber are equally suitable. There are a few exceptions. For instance, shredded wood is less good for livestock bedding as it reduces water infiltration. Chipping is also essential if growing for biomass heating, as the shredded material will block the feeding mechanisms and augers used in most systems.

The bigger the machine, the bigger the branches it will chip and the quicker it will do it. Buying a small chipper is likely to be a false economy with you spending days feeding small branches in and not managing to process the larger material. Secondhand commercial-scale, crane-fed machines can be picked up for £4,000 to £5,000 ($5,500 to $7,000). This would cope with branches up to about 20 centimetres (8 inches) in diameter. However, since you can hire a similar machine and operator for around £300 ($420) a day, you'd need to have quite a few days' worth of chipping to tip the balance in favour of owning a machine. Lack of storage space for unchipped wood while you wait to hire machinery is an important consideration and might be the deciding factor. It is also worth noting that if left for a long time in a field, weeds can grow up through the material and it can become enmeshed and more difficult to separate and feed into the chipper.

The really big chippers can devour whole trees and will get through huge amounts of wood very quickly. Make sure you have enough wood to keep them busy for a whole day before hiring.

HARVESTING WOOD FOR CHIPPING

In theory, the most time-efficient method of managing the chipping process is to have your chipper next to your cutting operation and feed the timber straight into the machine. You only have to move each branch once, and since the chip is more compact than branches, there will be fewer journeys to get it where you want. In practice, this may not always be possible, for instance if you are cutting with a chainsaw or where the ground is too wet to bring equipment in the winter. Even where you do have a chipper onsite, you are likely to need to move the wood a short distance from where it is cut to the chipper. Much of the specialist cutting machinery has grabbers or collectors integrated into the design, but if you are using your own kit and not operating on a small scale where you can lug small quantities of branches into a trailer by hand, consider investing in a front-mounted grab or forks for your tractor or telehandler.

MOVING AND TURNING THE CHIP

Some larger machinery is almost always essential for this task. Front loaders or buckets work fine, or a 360 digger can also do a great job. At a push, a link box at the back of the tractor can be used, though this is less efficient (it can work if you are then taking it out to spread from the link box). You can also use the bucket for turning the heaps. If you are not storing the chip on hard standing, you'll need to be careful about churning up the soil as you back in and out. A digger can be better in this respect; the tracked wheels cause less damage and you have greater flexibility of moving the arm and bucket. If you are planning to use a muck spreader or mulcher, make sure that your front loader is long enough to reach up in the machine. Whether you use a windrow turner (as described on page 61) or other simpler equipment is largely a matter of budget and scale, though specialised kit can improve quality of compost and decrease time spent turning.

SPREADING

Though it would be great to be able to chip straight into your spreader, in practice lining all the machines up ready is tricky. What is likely to happen is that the spreaders are sitting waiting for long periods while the chipper loads up, or the chipper is waiting for the spreader to come back before it can get to work. This can be expensive if you are hiring specialist equipment. It is usually better to chip into trailers, which are cheaper and usually available on the farm. You can then use an interim storage place from which you can take chip more efficiently when spreading.

The choice of equipment to use for spreading chip depends a lot on how you are using it.

Ramial chipped wood, used on field for soil health improvement, can be applied using any manure spreader. These are cheap and plentiful and available in a range of sizes. Rear discharge models are the most flexible and better able to deliver the chip accurately.

Case Study

Garth Clark checks the quality of chipped material from the Ecogreen shredder-mixer at Waddesdon Estate in Buckinghamshire, England.

On Waddesdon Estate in Buckinghamshire, England, the team is using woodchip on a wide scale and in a variety of ways. The land provides the space and resources to produce woodchip in significant quantity and the estate has invested in the machinery to make it financially sensible at scale.

Two hundred tonnes of woodchip is used in the biomass boiler each year to heat offices on the estate and the fine by-product material from the biomass is then mixed with other woodchip waste and manure from the estate's livestock to make compost. Garth Clark, estate director, has moved the management focus of the whole estate towards agroecology and he explains the estate's incorporation of woodchip in its methods:

Use of woodchip to make compost complements the other changes to our farming practices, such as increasing

Mixed woodchip and manure laid in rows along the field edge to mature at Waddesdon Estate.

herbal leys and the introduction of longer rotations, as the composted material rebuilds the biology of the soil and its carbon levels. We are already seeing improvements to our soils and we are now experimenting by adding charcoal into the composted woodchip mix, too, to further increase its quality and the soil's carbon sequestration properties. We are so convinced by the results achieved using the composted material we are producing that we are looking at how we can help other farmers to replicate our methods.

They can also be used for mulching paths. I have even seen growers modifying their spreaders by blocking off part of the outlet to apply mulch to paths between beds.

Side-discharge models designed for orchards are better if you are mulching rows of trees or shrubs. There is more detail about these in

Using a front loader to place mulch between trees at Eastbrook Farm. An effective, if somewhat imprecise, method.

chapter 6, on page 152. You can apply the mulch more exactly in both quantity and placement. They also tend to be narrower and so able to fit between rows of trees.

If you do not have any specialist spreading machinery and don't want to hire in, there are other options, but they are less effective. You can simply use a front loader to take the mulch where you want by the bucketload.

This can work well for rough mulching of paths or larger areas, but it is hard to do accurately if you are mulching trees. I have managed to mulch individual trees with a side-tipping dumper, though getting the right amount of mulch on each tree is impossible, particularly if the woodchip is partially composted and starting to stick together in clumps. You end up with either not enough or suddenly a whole load falls out the dumper, smothering the tree – definitely not a recommended method for small trees and shrubs.

For smaller areas, or for plantings that are inaccessible to larger machinery, a small tractor or quad bike with a trailer is still worth considering, particularly if you have a front loader to fill the trailer up. You can apply by hand very accurately to each tree. For large plantings, this does get very expensive.

CHAPTER FOUR

Woodchip for Plant Propagation

Raising good plants is fundamental to the success of any horticultural business. If you start with second-rate seedlings, you're likely to finish with second-rate crops. Most larger-scale growers either have dedicated propagators within their business or contract professional plant nurseries to raise them. In Europe, almost all plant production relies on peat as the main component for the growing substrate or propagation media. While there are clearly opportunities for large commercial companies producing growing media to replace peat ingredients in their mixes with woody alternatives, this chapter is aimed more at those growing on a smaller scale.

I should also issue a health warning at this stage. As this chapter unfolds you will see that it is possible to make a quality material from woodchip, but it is also easy to get it wrong. If you are relying on a consistent high-performing growing medium, take time to experiment and build your expertise in making it before moving your whole plant production system over to woodchip.

I'm going to look at three areas where there are opportunities to make better use of wood within plant-raising operations.

1. **Producing your own growing medium** from woodchip. It is possible to do this using simple techniques and minimal equipment to give a product that can match leading peat-based alternatives, even for raising seedlings.
2. Using woodchip to bring a **sustainable heat source into your early spring propagation**. Many tender plants need temperatures

Substrate Terminology

'Compost' is a term thrown about to describe a very wide range of products and materials. To avoid confusion, I shall clarify how I am going to use a range of terms in this chapter.

Compost. The result of a properly managed aerobic composting process, which sanitises the material through self-generated heat to give a high-quality, stable and biologically active material. True composts are frequently not suitable substrates for sowing seeds or growing young plants, because they can be too nutrient-rich.

Green compost. This is usually the result of large-scale processing of green 'waste' from gardens and or amenity horticulture. It does not include household waste. Arguably, composted woodchip could be called green compost; however, I will make a distinction in this book between them, which is mainly the purity of feedstocks and lack of greener materials, particular grass clippings.

Growing medium/substrate. This is a material specifically used for propagating and growing on plants. Growing media can be made from a wide range of materials (Table 4.2 on page 92 gives some information on the most common) and will have different properties depending on the requirements of the grower. In the UK, growing substrate is often labelled as compost, though it is very different from the true composts defined above.

above 18 degrees Celsius (65 degrees Fahrenheit) to germinate and grow to a point where they can be planted out in early summer. Growers can harness the heat from rotting woodchip to provide this warmth.

3. The role woodchip can play in nursery production as **a top dressing for pot-grown plants**.

What Is a Growing Medium?

A growing medium is a material used to sow seeds and root cuttings into or to pot up plants for growing on. It can be made from all sorts of things – usually organic materials like peat, coir and leafmould – but may include topsoil, as in the traditional John Innes mixes. John Innes was a nineteenth-century English property tycoon who left all his money to found a horticultural research institute. The institute developed a range of 'recipes' for growing media, with different recipes for seed, rooting cuttings, potting on and even ericaceous plants. The recipes were based on loam but included peat, sand and fertilisers. These recipes have become standard mixes in the UK and are still manufactured today.

By far the most common substrates sold today, though, are peat based. Peat is cheap, weed-free, uniform, effectively sterile and almost devoid of nutrient. It is acidic, but the pH can be adjusted during the manufacturing process to bring the product up to the desired pH value, which is usually around 5.5 to 7 (neutral), depending on the specification. This makes it an ideal base material for a growing medium. Historically, peat has been cheap to extract and easy to handle. Given all these benefits, why should we even be looking for alternatives?

The Problem with Peat

There are a couple of major environmental issues that are causing many people to completely reconsider our dependence on peat in horticulture. Firstly, peatlands are the world's biggest natural store of carbon; the International Union for Conservation of Nature (IUCN) estimates that the more than 3 million square kilometres sequester about 0.37 gigatonnes of carbon dioxide a year. If you took all the other types of vegetation on earth and combined them, they still wouldn't store as much carbon as peatlands. If you also consider that peatlands are less than 3 per cent of the total land area on earth, the carbon storage potential is even more astonishing.

Draining and harvesting peat has a major impact on this sequestering function. The IUCN has stated that as much as 6 per cent of manmade carbon dioxide emissions can be blamed on damaged peatlands. Not all of this is due to horticultural demand, of course. There are still significant volumes of peat burnt for heat and power.

Propagating without Substrate

Though slightly out of scope, it is worth noting that some crops are arguably best propagated with no substrate at all. Bareroot transplants (also known as peg plants) are sown in soil, often under cover in cooler climates. You just dig them up once they have four to six true leaves and transplant to their final positions. Not only does this method have the advantage of removing the cost of substrate, energy and plastic from the operation, but many growers find that it produces a hardier plant that can be more resistant to pest and disease attack. The other big advantage is the flexibility it offers around planting time. Modules and block-raised plants can run out of nutrient as they get larger; if you are unable to plant them out when you had intended (usually due to weather constraints) the plants will suffer. Peg plants will happily sit in the soil until you get the right conditions. Brassicas and leek are best suited to bareroot production.

However, where there are practical alternatives, we should be looking to eliminate peat use, and quickly.

Carbon is not the only problem with peat extraction, though. Peat bogs perform a range of ecological functions such as mitigating floods and keeping drinking water clean. Peatlands are also a vital habitat for many endangered species. Draining them can have devastating effects on biodiversity.

The 2008 publication 'Peatlands and Biodiversity' found that as well as individual sites being a habitat for numerous species, 'Peatlands support biodiversity far beyond their borders by regulating the hydrology and mesoclimate in adjacent areas and by providing temporary habitats for "dryland" species.' In addition, they state that, 'Peatlands are often the last remaining natural areas in degraded landscapes. Thus, they mitigate landscape fragmentation and support adaptation by providing habitats for endangered species and those displaced by climate change.'[1]

So, the case for moving away from peat in horticulture seems to be clear. However, many commercial growers are reluctant to switch fully, due in part to the historic unreliability of alternatives, but also the higher cost of peat-free alternatives. It is important, though, to better understand the true sustainability of alternative materials. A long-running project run by the UK government is due to release its results soon along with a calculator to enable new materials to be evaluated. As this research and debate continues, however, an increasing number of commercial alternatives, using bark, coir and even bracken and sheep's wool, are finding their way to market, and the quality of these products is also improving.

There have always been growers who have produced their own mixes onsite, sometimes entirely from home-produced ingredients or, more often, from a mix of bought-in and onsite resources. One major stumbling block has been the limit to which home-produced compost (which tends to come mostly from vegetable waste) can be added to a mix. The problem here is the high levels of nutrients and high pH of green compost, which can inhibit seed germination. The sustainability charity WRAP recommends is that no more than 10 per cent of a seed mix should be green or homemade compost, though this can be higher for bedding and potting composts.[2] It is worth mentioning that for many crops it doesn't seem to make that much difference, particularly those with larger seeds.

In the UK, there is a legal standard for making green compost. The Compost Certification Scheme's compost quality standard, the Publicly Available Specification for Composted Materials (PAS 100), lays out certain requirements for such things as ingredients and composting process to guarantee a quality product. Unfortunately, this guaranteed quality has not historically been high enough, with levels of plastic contamination a particular concern. A review of the standard in 2018 intended to improve the overall quality, but concern about Green Compost is one reason to consider growing or sourcing a woodchip product with lower risk from contaminants.

Peat for Blocking

There is a major barrier for growers using blocks instead of modules or trays for raising seedlings. Blocks are self-holding cubes of substrate with a small depression in the top into which the seed is placed. They

Table 4.1. Suggested Rates of Use of PAS 100 Green Compost in Growing Media

Use	Percentage Range of Composted Green Material
Seed mixes	5–10%
Bedding mixes	20–25%
Pot plants	20–25%
Nursery stock (general, excluding ericaceous species)	30–35%
Nursery stock (vigorous, excluding ericaceous species)	35–50%
Multipurpose growing media	20–40%

Note: These suggestions are for material mixed with low nutrient, low conductivity material to make growing media. If using a high conductivity material, then use the lowest rate.

Source: 'Guidelines for the Specification of Quality Compost for Use in Growing Media', WRAP (February 2014), https://wrap.org.uk/resources/guide/guidelines-quality-compost-use-growing-media.

are quicker to plant; you simply lift the block and place it in the planting hole (or for some crops you can literally pop it on the surface and, provided the weather is not too hot and dry, the roots will find their way from the bottom of the block into the soil). Advocates of blocks swear that they produce a better seedling and that you don't get the root binding problems that can occur in modules. Instead of hitting the module or pot wall and growing round and round, when a root reaches the edge of the block it finds air and so simply stops growing. This means that when you plant the block there is theoretically less of a check in growth that you can sometimes get from seedlings that have been in their modules for too long. Currently, peat seems to be the only material that holds together well enough in a blocking machine. Though some mixes are now produced with reduced peat, work is still needed to develop a totally peat-free blocking compost for growers.

Leafmould is a great material to dilute the effects of green compost. However, it is almost impossible for any but the home gardener to find enough to be useful. Even assuming a good supply of leaves, few are willing or able to spend the time collecting and transporting them. Woody material can have similar properties: a high carbon to nitrogen ratio and a bulky, light structure. What hasn't been fully explored is the potential to produce growing media from pure composted woodchip. Though in most parts of the world there is an ample supply of waste wood material, it seems no one has considered the potential to turn it into a propagation material. One grower in the UK has succeeded in doing exactly this.

Tolly's Homemade Woodchip Compost

Iain 'Tolly' Tolhurst, owner of Tolhurst Organic, has been making compost from woodchip since 2007. Here are some of the lessons he has learnt from this experience.

Tolly has been growing stockfree organic – in other words, with no animal inputs – at the Hardwick Estate in Berkshire, England, for thirty-three years using a range of organic techniques to improve his soil but also reduce his inputs. He was an early proponent of green

Woodchip broken down after one year. Sieving easily removes the larger particles.

manures and undersowing in horticulture in the UK and is regarded as one of the most pioneering and influential growers. Like many growers, he is a perfectionist who doesn't believe in relying on others to deliver the quality he needs. However (also like most growers), he has a lot on his plate and over the years has found it easier to buy compost in and focus his attention on other parts of the business. A bad batch of compost persuaded him to look again at producing his own material onsite. Prior to the BSE (bovine spongiform encephalopathy) crisis in the UK, he had removed fish, blood and bone from propagation substrates, as he considered them unethical as an input.

He had been playing around with woodchip as an additive for soil health (see chapter 5) and, being one of life's rebels, took no notice of the conventional wisdom that said you can't compost woodchip or use it in mixes due to the risk of nitrogen lock-up. He observed that the composted woodchip coming out of his process, if sieved, could work well for propagation, and so he started experimenting. Early successes convinced him that there was potential, so he decided to broaden his experiments and get other growers involved.

This led to a field lab as part of the Soil Association's Innovative Farmers programme. This programme brings groups of farmers and growers together to tackle a research question on their farms, but brings in a researcher to help them design the trial, in the hope that these farmer-led innovations will more likely show a clearly defined and repeatable result. Moreover, by working together the farmers tend to carry through more often with the test or trial, as working on your own runs the risk that the trial gets dropped as the season gets busy. The outreach part of the programme also ensures that positive results get shared beyond the farms that took part.

Tolly's field lab compared his homemade woodchip compost to an industry standard peat-based commercial substrate. He also added some biochar; he had been mixing in perlite originally to lighten the mix, but was keen to find a more sustainable alternative.

Details of the Trial

Both the woodchip and peat composts were tested with and without enriched biochar in a randomised trial with cabbage and leek seedlings. As you can see from the charts on pages 86–87, there were some noticeable differences between the composts, in particular the levels of nitrogen in the woodchip compost. Many of the effects of the biochar may have been caused by the enrichments in the biochar product used in the trial – seaweed and worm casts, for instance – rather than in the carbon part of the material.

Data on the germination, growth, and final health and yields of the plants was collected. Early, quicker growth on the cabbages in peat seemed to even out over time and show no difference in final performance. Overall, there was no statistical difference between the different treatments, with one exception. At the time of harvest, the leeks were beginning to show signs of leek rust. The peat compost without biochar showed significantly more infection than the other three treatments.

Financial Details of Compost

Homemade woodchip compost might be the technical answer to all our prayers, but if it doesn't stack up financially growers won't move away from their current products. Tolly has costed his process, looking at the cost of materials (free), the hire of necessary machinery (though many farms will have a suitable option onsite already), and labour costs to turn and process the materials to get the final product. He also added in the cost of a small quantity of perlite to lighten the material. He found that it was no more expensive for him to make his own than to buy in the best-quality imported substrate.

Tolly's Summary of the Trial

'The results of the trial, from my perspective, were very interesting, as it was clear that our material – which is low-tech, reproducible on farm and produced from predominantly local materials – was a very good match against a peat-based imported product. The addition of biochar in the case of leeks did show a slight improvement of plant quality and is worthy of further investigation and trials.'

What You Need to Make Your Own Woodchip Propagation Substrate

Key to success is the right raw material. In many areas of the UK, it is not difficult to get hold of woodchip from tree surgeons. Operators pay a gate fee at recycling centres or landfill sites to dispose of

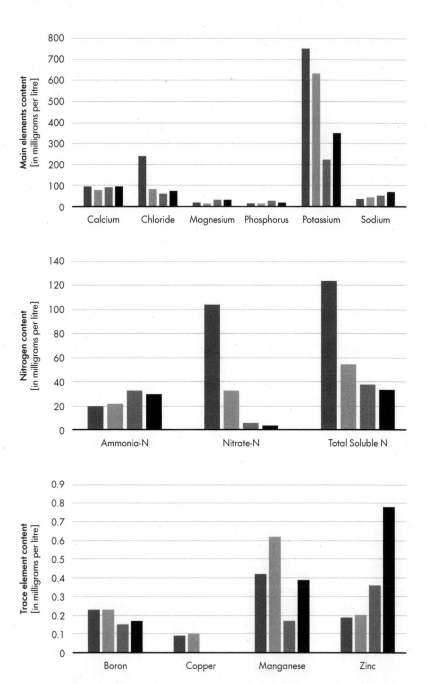

Analysis of woodchip compost compared to leading peat-based brand. *Source:* Innovative Farmers 'Growing with Peat-Free Woodchip Compost.' Field Lab report.

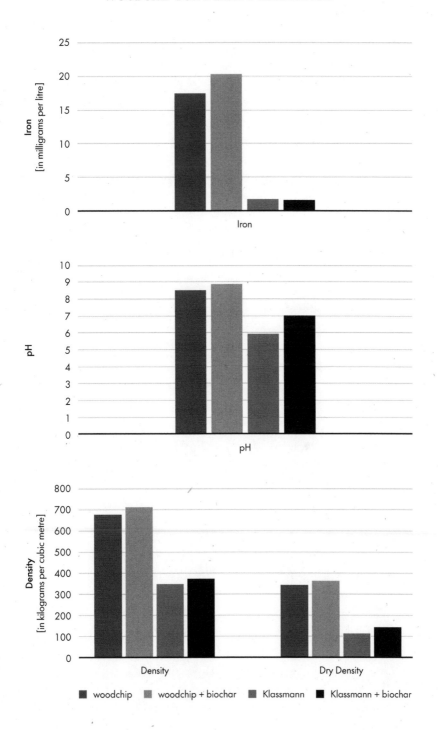

woodchip. Some arborist and recycling businesses have recognised the value of woodchip and are making their own products, which they sell. For those that don't, being able to offload it for free with a friendly farmer is a bonus. As I mentioned in chapter 2, there are even websites that match up tree surgeons with material to dispose of with people looking for free woodchip; search for Arbtalk if you are in the UK and Chip Drop for in the US.

The risk with accepting free woodchip, though, is quality. Tolly has a good relationship with his supplier and is picky. His tree surgeon knows the type of woodchip he wants. Access is relatively easy to Tolly's site, and this makes a difference, too. Remote farms, or those with very long bumpy drives, might put a potential supplier off. At Eastbrook Farm we are lucky to have a reliable tree surgeon living in the next village. He passes by on his way home, so we are his favourite site to offload his chip.

The eventual aim for many growers is to be able to supply your own woodchip. You have greater control over the species mix and quality and can manage on your own terms. Whether you are growing your own or getting an outside supply, here are a few rules:

- Don't include anything too big; ideally, use material less than 7 centimetres (about 3 inches) in diameter. The reason for this is to give you sufficient nitrogen content from the living tissue in the bark and buds, which allows for a quick breakdown of material to prevent nutrient lock-up in the final product.
- Some conifer and evergreen is okay, but not too much – only include up to 30 per cent.
- Ideally, you want a mix of leafy and woody waste, though in practice this is more determined by the time of year. In winter it tends to be woodier, while in spring and summer there is more green content.
- Ensure there are no contaminants mixed in. These can take a number of forms, from plastic to building rubble, big logs or even chainsaw helmets and pruning saws. If your supplier cannot provide clean material, stop using them.
- Disease is a potential risk, especially as one of the main reasons that tree arborists are asked to fell trees is because they are diseased and dying. This risk is very slight in propagation as the diseases of trees are unlikely to be the same that affect vegetable

seedlings. In addition, the composting process should reduce most diseases to the point of irrelevance.

Method

Tolly gets two or three loads of woodchip a week; each load is added to the end of his windrow. He has two rows on the go at any one time. The first is completed and just gets rowed up and turned, while the second is the live one that is being constantly added to (though this, too, will be worked on during the season). The rows can be as long as needed to accommodate the material.

Woodchip top dressing a container plant.

He receives approximately 500 cubic metres (650 cubic yards) of chips per year, which eventually composts down to about 250 cubic metres (325 cubic yards). He turns the heap every three months or so with a small digger that he bought specifically for this purpose. (At least, that is what he claims; actually he just likes playing on diggers.) So, he aims to turn the rows four times during a year. You can do it more frequently, and that might speed up the process a little, but Tolly reckons four times gives optimum performance for his time and effort.

The heap tends to go semi-dormant in the summer, because, unlike in a traditional compost heap where you would be relying on bacteria and a range of other organisms to break down the greener material, woodchip needs fungi. Fungi shut down in hot weather. Tolly has tried irrigating the heap to try and get it working again in hotter, drier conditions but found that this had little effect. The heap did not absorb the water, and it just ran through and out onto the ground under the heap. Rather than stressing about this, he has now accepted that the breakdown of the heap takes a holiday during the summer and starts again once the cooler autumn weather sets in. Tolhurst Organic is in Berkshire, England, where the annual rainfall is around 550 millimetres (21 inches) and summers can be very dry. In a wetter climate, the semi-dormant phase may be shorter or even not happen at all. Very high levels of rainfall could make the materials too wet, particularly if the material has completed the composting process. Using a cover at certain times of year would reduce this risk.

In very hot and dry climates, it may be necessary to cover the heap and irrigate it to provide suitable conditions for fungi to thrive. David Johnson's cold composting system (the Johnson-Su Bioreactor) in New Mexico does exactly that, with sprinkle irrigation from the top and a moisture-holding top layer.

Tolly does not cover the heap. This is partly for management reasons, as it would make it more time-consuming and fiddly if he had to uncover the heap every time he needed to turn it. He also wants the moisture from rain and dew to enter the heap to aid fungal decomposition. He has considered that covering it during the hottest weather might help to keep it cool and therefore breaking down, however, on balance, he has opted for patience as the simpler if slower path.

The only exception to this is the smaller pile that he composts further for his propagation medium. Since this is left for longer and needs to be weed-free, it is vital to keep this covered for the final six months of composting. In total, from receiving the raw woodchip to final composted material, at least eighteen months is needed to ensure material is suitable as a propagation substrate.

Many larger composting sites have installed concrete pads. In many cases, this is a regulatory requirement, with bunded runoff capacity; in other words, with a self-contained area that catches any spillages to prevent water pollution. This is, of course, important if you are composting manure or other materials with a high nitrogen content. However, for the woodchip this seems not to be too important. As mentioned above, in the event of very heavy rain it just runs through the heap without taking anything with it, and in light rain the moisture is absorbed into the heap. The woodchip appears to hold onto any nutrients pretty strongly. If the heap were left for longer, it is likely that nutrients would begin to leach, but by that time Tolly has spread it onto the land and any soluble nutrients will have been taken up by the green manure crops.

Tolly believes that siting the woodchip piles directly onto the soil surface is actually a benefit. It seems to speed up fungal breakdown, since there is already fungal activity in the soil. The heap is also able to draw moisture from the soil, so that in the event of hot weather the bottom of the pile continues to be cooler and more active. Siting on soil does, of course, create small operational challenges in terms of machinery; however, due to the low-tech nature of the site and the flexibility of the system, it is rare that Tolly is not able to get out frequently enough to ensure the process works well. The use of a small digger (swing shovel type on tracks) means that access is possible under any conditions and turning speed is high. Front loaders on tractors are generally unsuitable as they require lots of room to operate, are slow, and do not allow for the selection and mixing of materials within the turning process.

As you are likely to be receiving woodchip and making compost every year, you should always have a good supply of fresh material to hand. However, if you wanted just to make larger one-off batches of compost for propagation, it seems that it does have a long shelf life. Tolly has trialled some four-year-old material that had been stored in sacks. It had dried out so needed rewetting, but otherwise performance was no different to younger substrate of the same mix.

Smaller-scale operations, and even home gardeners, could use a similar method. The process relies on fungi, not bacteria, and so is likely to be successful in small piles. Small heaps in a compost bin or even using dumpy bags would work. Doing it in this way means that you will not need to turn, but it is worth covering as the heap will be there for longer than those in larger, more managed piles.

Audrey Litterick is one of the authors of WRAP's 'Guidelines for the Specification of Quality Compost for Use in Growing Media'. She works with growers in the UK and warns that, 'Making your own growing media is a difficult thing to do, and even experienced growers make big mistakes when creating their own. Making growing media of sufficient quality for germinating seeds and producing young plants is particularly difficult. The reason most people fail to make good seed propagation compost is that the media they make are either far too salty and nutrient-rich, or they are structurally all wrong (hold too much or too little water), or they contain damping-off pathogens. Indeed, they may well have all these problems.'

Table 4.2. Assessment of a Range of Growing Media

Material	Pros	Cons
Soil-based commercial product	• Holds nutrient well	• Heavy to transport and work with
Soil from own farm or garden	• Holds nutrient well • Plants grow in same medium as final planting place	• Weedy; quality depends on quality of soil • Sets hard and no structure • Unreliable • Physical properties of soil change when you put it in containers • Disease often a problem • Poor drainage
Peat	• Sterile and low in nutrient – allows nutrient control with additional feeding • Uniform • Light and easy to work with	• Difficult to rewet if it dries out • Biologically inactive
Coir	• Uniform • Free draining • Light and easy to work with	• Doesn't hold onto nutrient so needs more feeding
PAS 100 green compost	• Readily available	• Variable quality • High in nutrient for seed propagation • Usually immature so needs finishing • NOT suitable for seedling propagation
Bark	• Readily available	• Needs very thorough composting • Mostly not suitable for home production
Woodchip	• Readily available • Easy to process • Weed-free if processed cleanly • Good structure	• Takes time to process • Quality of feedstocks variable if not producing on farm • Small risk of disease

Environmental Impact	Cost
• Where the soil comes from will have an impact	High
• None	Free if you use your own soil
• Harvest releases carbon dioxide • Many peat sites are unique habitats	Low
• Coir is a waste product, but coconut growing can be intensive • Transport across globe also an issue	Medium
• Little, if well made, though plastic, glass and herbicide contamination is an issue	Low
• Good use of 'waste' resource from forestry mills	Low
• Low	Low

Table 4.2. *continued*

Material	Pros	Cons
Additional ingredients in small percentages		
Perlite	• Improves drainage and aeration of medium	• Light and can blow away • Dust from it can cause breathing problems
Vermiculite	• Similar to perlite, but holds more moisture	• Can hold too much moisture
Biochar	• Holds onto lots of moisture and nutrients, reducing the need for watering • Potentially also stimulates germination and initial growth	• Can hold too much moisture • If used without nutrient amendments it can cause deficiencies

In her experience, after years of producing inconsistent media, some growing media companies are now getting it right, and that 'there are commercially available peat-free growing media which are as good as peat-based types. However, if these media do contain composts of any sort (and they usually do not), then the composts are usually heavily diluted with either sterile coir, perlite or vermiculite (all of which have sustainability issues) or sterile wood fiber (which to my knowledge is unobtainable in small quantities)'. True composts, where used, usually constitute between around 10 to 30 per cent, at most, of a growing medium. The most recent report into UK growing media shows that wood-based materials have been increasingly used as a component in commercial products, though they still only account for around 20 per cent of retail and 10 per cent of professional ingredients used for substrates.[3]

It is easy to think that just because your homemade compost looks like a commercial product it will perform like one, but, in fact, this is often not the case. I say this not to put you off. I strongly believe from seeing Tolly's system that more of us could be making our own, but it's wise to caution that there are many ways to ruin a good compost.

Environmental Impact	Cost
• High – mined and takes a lot of energy to produce	High
• High – mined and takes a lot of energy to produce • Lower carbon footprint than perlite	High – higher than perlite
• Depends on feedstock and where made • One-third of energy potentially lost in manufacture. However, long-lasting carbon sequestered in soil when planted out	High

Woodchip Hotbeds

The concept of the hotbed is not new. We know the Romans harnessed the heat released by decomposing organic matter to extend the growing season, and in eighteenth-century England growing pineapples in special beds heated with large quantities of manure was common amongst the ultra-rich. Records from 1755 show the famous naturalist Gilbert White using them for melons and cucumbers, too. A hotbed is effectively fresh organic matter that you pile into a frame. You can then place your trays or pots of plants on top or near to this frame, and the heat keeps the soil and air warm. Although you could place the plants directly onto the composting organic matter, it is more usual to place a board or wire mesh of some sort over the decomposing material. This gives you a flat surface to sit the trays on. Traditionally, gardeners use manure of some sort, usually mixed with straw. This gives a good initial heat and then gradually cools down. To get continual heated production, you would need to keep replacing the rotted manure with fresh. Many growers just use this system in early spring to protect their seedlings.

Fred Bonestroo, the grower at Duchy Home Farm near Tetbury in Gloucestershire, England, was inspired by Tolly's woodchip experiments and saw the heat that was given off by the rotting woodchip. He designed a modified hotbed that used freshly chipped woodchip. It is important that the wood used is relatively small-diameter branches, as you need good nitrogen content in the woodchip to ensure quick and hot breakdown. In traditional hotbed systems, this nitrogen is provided by the manure.

Fred noticed that although his hotbed did not get quite so hot as a traditional manure one, it did not cool so quickly. This meant that he was able to provide a more even temperature for his seedlings and that heat was generated for longer than in manure hotbeds. He can keep his hotbed going from February until April. The temperature never falls below 21 degrees Celsius (70 degrees Fahrenheit). He has raised chillies, sweet peppers and tomatoes using this system.

The final benefit, of course, is that while fresh and rotting manures can be a little harsh on the nose, particularly in the confined space of a tunnel, decomposing woodchip is a wonderful scent that transports you into a forest as you work.

Size does matter. You need enough bulk for the woodchip to get to a high enough core temperature. Ed Sweetman, another grower

Fred's woodchip pallet-frame hotbox at Duchy Home Farm.

who has taken on Fred's model for this system, makes his hotbed three pallets long by one deep. This takes around 3 cubic metres (4 cubic yards) of chip to fill it and can fit fourteen 216 module trays on it. He estimated that it took about four hours to load each time, and getting the material dumped close to the tunnel can be tricky.

Fred uses a see-through acrylic sheet on one side to allow some sunlight in, which warms it a bit in early spring; the other side is covered in wood. He puts a plastic cover over the top of the hotbox to maintain moisture and prevent the woodchip drawing water from the seedling trays. This is then covered with clear plastic resting on hoops.

Wood Type

Selecting the right age of wood for making a hotbed is crucial, just as it is when making your own propagation substrate. The wood should contain enough nitrogen to begin a quick breakdown. Ramial chipped wood is the best, and it should be freshly chipped. If left too long after chipping before making the hotbed, you will have lost some of your potential heat. If you can only get hold of older wood, it might be worth mixing this with a nitrogen source (for instance green garden materials, pack house waste or even fresh manure). If the chip has got very dry, an initial watering as you load can get it going more quickly.

The ideal scenario in UK conditions, if you want to start propagation at the start of February, is to cut and chip the wood in the last week of January and make your hotbed immediately. You should have enough heat to see you through till at least the end of April.

This method is so effective that there are times when the hotbed will give out too much

Fred lifting a Perspex cover on the woodchip hotbox.

heat and risk burning the roots of your plants or pushing growth too hard. In this case, you can simply raise them up off the surface of the woodchip by placing an empty plug tray underneath to allow some air movement.

Once you have finished propagating, you can leave the woodchip in situ till later in the year when things are quieter. There is a risk, though, that you will attract pests into your propagation unit (especially mice and rats that might see it as an attractive home). Whether you dismantle it as soon as you have finished using it or wait till later in the season, the woodchip should be well enough rotted down to add immediately to your soil, either as mulch or scattered thinly to improve soil health. If it is not fully rotted, do not dig the chip into the soil as this can cause nitrogen lock-up. Fred tends to leave his until very early in the following spring when the woodchip has rotted well down before removing and spreading it onto his no-dig tunnel beds. This also ties in with him preparing the next year's hotbed.

Woodchip as a Top Dressing for Pot-Grown Plants

Hand weeding pots is one of the major costs in producing nursery stock. If you don't keep your pots weed-free you increase watering costs and can affect plant growth and health. Even if you are using a sterile substrate like peat, as soon as you stand the plants out there are weed seeds flying around in the air that will land in your pots and quickly take hold if not controlled. There are various ways to reduce weeds without hand weeding.

If you're not producing organically, herbicides are an option. Usually, pre-emergence herbicides are used, which prevent weed seeds germinating. However, post-emergence options are available as well, though they are more costly to apply. Herbicides are becoming increasingly unpopular with consumers and may influence the biology of the substrate, and therefore the health of the plant.

The other alternative is a physical barrier that blocks light from the weed seeds and prevents germination. There are a range of mulch mats available manufactured from materials such as jute, coir and wool. They do a great job of weed control for plants with a single hard stem, but don't work so well for herbaceous or soft-stemmed plants. They can also make watering a bit tricky if you don't have drip irrigation systems as the water can run off the mat rather than being absorbed through. Also, they are not cheap.

Using woodchip to top-mulch pots is an option for growers of all sizes; indeed, many of the largest ornamental nurseries recognise the benefits of this method over chemical control or hand weeding. But it is also ideal for home gardeners.

There are a few things you need to think about, though, before rushing out and smothering all your pots in woodchip. Unlike mulching open soil, where there is a large reserve of soil life and nutrients, in a container (and especially in small pots) it is easier to significantly change the growing environment. You might well be adding 20 per cent by volume with your mulching material. If, for instance, this mulch has a very high or low pH, this might affect plant growth. There is also the risk of spillage. Woodchip is a lightweight material that can easily fall out of the pot if it gets knocked or blown over. Birds also like scratching around on the surface of larger pots. Weigh up that extra time spent sweeping up woodchip against time saved in weeding.

The age and the size of the woodchip needs to suit pot size. Applying fresh large chip to small pots is likely to cause nitrogen lock-up and check the plant health. However big containers with mature plants might cope just fine. For smaller pots use smaller chip; if you can use partially composted chip that might also help, though if this composting has happened in the open air there is a risk of bringing in weed seeds. Usually, if there is still a woody structure to the mulch the weeds will struggle to germinate even if there is seed present. Pine bark mulches are one of the more effective ways of controlling weeds in pot-grown plants. This is partly because they contain a range of naturally occurring chemicals that inhibit germination of weed seedlings.

Woodchip can work well for large permanently potted plants. I remember visiting a walled garden that had quite a few big topiary plants in containers. Each year the level of the substrate in the containers would drop by a few centimetres as the plants took nutrients to grow. They initially had topped this up with their homemade compost but found that this disappeared too quickly, so they experimented with a thicker layer of woodchip and found this worked far better. Each year they added the new mulch on top of the half-decayed layer from the previous year. The pots were full of worms and retained moisture better, meaning less watering.

In extreme temperatures, potted plants are at risk. Ice and snow, fierce sun and even strong winds will all put your plants at risk, particularly when the plants are very young. A layer of woodchip on top of the growing medium will help mitigate some of the effects of these weather conditions, protecting the plant from frost or drying out and keeping the actual temperature of the substrate a little higher or lower by sheltering and shading.[4]

The final advantage to woodchip mulching your pots is aesthetic. Though this may seem a minor concern, as a commercial nursery if your plants look great when set off by a lovely even dark bark mulch, you are likely to sell more. If they also stay weed-free for longer at point of sale, your return and wastage costs will almost certainly be lower.

A typical mix of woodchip from a tree surgeon includes a wide range of material, deciduous and coniferous, and different-sized chip.

Single-species willow woodchip; even-sized chip that fit well through the forage wagon.

A browsing block of willow and poplar at Eastbrook Farm, mulched with willow woodchip.

A woodchip pile placed at the end of a tree row, ready for mulching.

Coprinopsis fungi growing on woodchip mulch.

Strawberry transplants raised in woodchip substrate looking vibrant and healthy.

Woodchip and some garden waste laid out in windrow.

Chard grown under cover with woodchip mulch in tunnels at Duchy Home Farm.

Ramial chipped wood (RCW) on soil at Down Farm, showing what the RCW looks like when spread at the Organic Research Centre's recommended rates. Photo courtesy of Sally Westaway.

Worms find their way into the heap, even when there is still a lot of unrotted wood.

Willows mulched and unmulched, all planted at the same time in 2018, showing the striking effect of a deep mulch in a drought year.

Creeping thistle beginning to show through woodchip mulch.

Fred Bonestroo even uses the composting piles of woodchip to grow crops in; here is some fine-looking squash growing well in one of his piles.

CHAPTER FIVE

Woodchip as
Soil Amendment

Perhaps the biggest barrier to better and more informed use of woodchip for soil health has been the overstated problem of nitrogen 'lock-up' or 'robbing'; where wood sucks up nitrogen from the soil, leaving none for the plants and leading to unhealthy crops. This has led to a general perception that adding woodchip to soil is dangerous and should be avoided. As we have seen, the fungi that break wood down *do* need some nitrogen to thrive. In the absence of other sources, they will take that nitrogen from the soil. Using the wrong wood in the wrong way will cause problems, and in the worst cases cause quite serious long-term soil health problems. However, a basic understanding of how wood rots easily helps to avoid these problems so we can unlock the potential of using woodchip to build our soil health. The single biggest thing we can do to reduce the risk of nitrogen lock-up is not to incorporate the chip into the soil but to spread it on the soil surface. Let's look in a bit more detail at how nitrogen moves about in the soil and its interaction with rotting woody material.

The Nitrogen Challenge

Managing nitrogen is one of the most important challenges for growers. Despite being the most abundant element in our atmosphere, where it exists in its gaseous form, N_2, it is notoriously tricky to get the right amount of it at the right time. Getting it from the air requires a lot of energy; the strong bonds holding the two nitrogen atoms together have to be forced apart to allow them to form new bonds with oxygen, thus producing nitrogen dioxide (NO_2). This process is called fixing. Lightning produces sufficient energy, and some nitrogen is fixed this way and brought to earth as NO_2 dissolved in rain. Most, however, is fixed by bacteria. These specialist bacteria are plentiful in the soil and do more than 90 per cent of all nitrogen fixing. Some are free living and others form a symbiotic relationship with some plant species, especially legumes. This is the reason so many farmers include nitrogen-fixing plants in their systems. In a traditional farming rotation, a fertility building phase, consisting of a mix of plants usually including clovers or other legumes, grows the nutrients, which are then exploited during the cropping phases. There is an added challenge, however. Though there may be plenty of total nitrogen in the soil, if the nitrogen is in organic form it won't necessarily be able to be taken up by plants. It needs to be 'mineralised', meaning converted to a soluble mineral form that plant roots can absorb in the soil water. The final conundrum is that its solubility, the very attribute that makes it usable, also means that it is very easily lost from the system – for instance, when water is lost from the soil in heavy rains.

Artificial nitrogen fertiliser has allowed farmers to bypass some of these natural processes and provide plants with an instant fix, but at a considerable price. There is a high environmental cost both in the carbon footprint of creating the fertiliser but also in the risk of runoff on application into watercourses, wetlands and the sea. The World Resources Institute estimates that 20 per cent of nitrogen fertilisers applied are lost through runoff and leaching. While there are ways to reduce this, for instance through precision applications, timing, catch crops, or the use of biochar, relying entirely on artificial sources of nitrogen is still an inexact and potentially wasteful approach.[1] There is also evidence that sustained use of nitrogen fertiliser can deplete soil carbon and nitrogen. The theory is nitrogen fertiliser 'depletes soil organic matter by promoting microbial C [carbon] utilization and N [nitrogen] mineralization.'[2] In other words, if you feed nitrogen to the microbes this speeds them up so they use more carbon, but if you don't replace that carbon with more your soil will gradually lose its organic matter. In addition, the ready availability of artificial nitrogen inhibits both microbial fixation and fungal exchange of nitrogen via mycorrhizal associations.

The key to our plants getting the amount of nitrogen they need when they need it is in getting the soil biology right. Mineralised forms of nitrogen are released as a by-product of life cycles, when plants, microbes and other animals die. A biologically active soil will release a constant trickle of nitrogen into our system to feed our plants. The key to achieving this aim is to feed the soil. As we have seen, cover crops with legumes and other plants will add fertility, but we can also add organic matter in the form of compost and, of course, woodchip.

Understanding this solubility of nitrogen in our system helps us figure out how to manage this risk of loss and lock-up. Soil water is subject to the rules of gravity like the rest of us, and mostly flows downwards. There are, though, other forces at work, and water can travel against gravity through capillary action, and at a cellular level through osmosis, but these forces tend to work in over a relatively short space. If you place nitrogen-hungry woodchip on the soil surface, it will only be able to suck up water, containing the dissolved nitrogen (which itself is only a tiny percentage of total soil nitrogen), from the surface layer of soil. Plants with very shallow roots, in other words mostly seedlings, will likely be affected. Indeed, this may be one of the mechanisms that helps mulches be an effective weed control, by preventing weed seedling establishment. Larger plants with deeper roots will be unaffected by this shallow, and temporary, nitrogen lock-up. After a couple of years, the woodchip will have rotted enough to release that nitrogen back into the soil in any case. A study by E.M. Miller and T.R. Seastedt looking at spreading woodchip in a post-thinning forest situation found no difference in availability of nitrogen in the first two years following the application of the woodchip, but increased nitrogen availability in year three.[3]

Rain falling on the woodchip will provide some of the nitrogen needed to feed the fungi to break it down, and in heavy downpours much of that rain will pass through the mulch to the soil surface, helping to alleviate any robbing effect.

When Not to Use Uncomposted Woodchip

Firstly, let us look at situations where you would not want to use fresh uncomposted woodchip (with the possible exception of ramial chipped wood, which I will come onto in due course). Plants that need nitrogen in the top layer of soil are those with tiny or shallow roots. The most obvious of these are recently germinated seedlings, which need immediate nutrient availability in the surface soil layer to continue growing once the stored food from the seed is used up. Small seeds are likely to be even more vulnerable as they will have smaller reserves. So, clearly avoid using uncomposted chip in propagation composts or in most vegetable and cereal crops where you are directly sowing into the top couple of centimetres of soil. Many flowers would also fall into this category.

There are other plants with shallow rooting systems that might be affected. Plants on shallow soil where there are fewer deep reserves of nitrogen are likely to be at higher risk. Similarly, rhizomatous plants, such as raspberries, or even young trees and shrubs with smaller root systems, may be unable to send their roots so deep and might be more susceptible to lower surface nitrogen levels. This is largely compensated for by extensive mycorrhizal networks in long-lived perennials.

Don't Dig It In

The evidence seems to suggest that any nitrogen robbing occurs locally at the point of contact between the woodchip and soil. There are claims that the chip will take nitrogen from up to 10 millimetres (½ inch) away; though I have not found any studies to back this up, it seems plausible from my experience. Placing the chip on the surface reduces the area of that surface contact. However, digging in the chip hugely increases those points of union and therefore the risk of nitrogen lock-up. The smaller the size of the chips, the greater that risk too, because of greatly increased surface area. The most damaging option would be to dig in uncomposted wood sawdust. Even with composted material I would favour adding woodchip to the surface of the soil. If you have healthy soil, the worms and other organisms will quickly incorporate it for you.

Adding Nitrogen

Though I would argue that the evidence mostly suggests there is generally little need to complement your woodchip mulches with a high-nitrogen material, you might in some circumstances decide to do so. We are experimenting with manure–woodchip mixes on our raspberries, as I feel they are not growing as well as I would have expected with a pure woodchip mulch. You might also consider adding nitrogen if you are spreading fresh chip on the surface, but intend to cultivate the soil later, perhaps for weed control. As you cultivate, the chip will become incorporated with the upper soil layer and so take on some more of the soil nitrogen.

As with most soil science, we are still learning. Multiple complex interactions are happening, dependent on soil type, climate and crops. This is shown in one study comparing various mulches, trying to take a holistic evaluation of soil health, tree performance and weed control.[4] What they found was that:

Wood chip mulch resulted in excellent tree growth ... In contrast, tree leaf nitrogen was lower than the other treatments ... Increased compost amendment rates led to higher leaf nutrient levels but not enough to meet levels considered desirable for young non fruit bearing trees, indicating that tree nitrogen needs were not met even at the high nitrogen rate. The wood chip mulch treatment resulted in abundant soil moisture, which may have contributed to the observed desirable tree growth; yet, this may also have contributed to nitrogen loss through denitrification or leaching mechanisms, as suggested by the low available soil nitrogen in summers. In concert, these findings demonstrate that wood chip mulch is not a suitable strategy for orchard floor management during tree establishment unless additional steps are taken to manage nitrogen supply.

This is fascinating as it goes against what might be our observational conclusion that 'these trees are growing great with the mulch'. The answer might not be, however, to abandon all the other benefits this study found of increased soil moisture and good growth, but rather to add a higher-nitrogen material into the woodchip prior to mulching. As we saw in the Miller and Seastedt study, by year three it is likely that there will be an increase in soil nitrogen from the mulches, so we can potentially compensate for this slight insufficiency of nitrogen during the initial establishment. For most growers, it is possible that the levels are not enough to worry about and might just result in slightly lower yields in the first year or two. For larger commercial fruit growers,

Table 5.1. Comparison of Carbon:Nitrogen Ratio for a Range of Materials

Material	Carbon:Nitrogen Ratio
Grass clippings	9–25:1
Leaves	40–80:1
Sawdust	200–750:1
Woodchips – hardwood	451–819:1
Woodchips – softwood	212–1,313:1
Coffee grounds	20:1
Cow manure	11–30:1
Poultry manure	12–15:1
Ramial chipped wood	70–150:1
Straw	48–150:1

however, this could make a significant difference to the profitability of a new planting.

Table 5.1 contains some examples of the carbon to nitrogen levels for some typical amendment materials.[5] As you can see, there is considerable variation in some due to the variability of the source.

The Ratio of Fungi:Bacteria in Soil

When soil is first formed from bedrock, bacteria are the first colonisers to move in. Bacteria, unlike most other organisms, can eat rock.[6] Once they have started to break down the rock surface, organisms like protozoa, nematodes and other microorganisms gradually arrive. Fungi also mine rock for minerals, but it appears that unweathered bedrock does not have enough of all the other things that they need to be able to get a foothold.[7] As the depth of soil increases, the proportion of fungi increases. Ephemeral and annual plants, including mosses, which trap dust and debris further enhancing soil development, are generally the first plants to take advantage of this newly formed soil, while shrubby plants and trees tend to require deeper soils and take longer to establish. Once a soil gets older, and supports a larger proportion of woody plants, so, too, the fungal percentage increases. Thus, from a soil initially dominated by

100 per cent bacteria, the soil develops until the ground in a mature forest is likely to have a bacterial to fungal ratio of 1:100, measured by biomass. The best ratio for any given plant depends on which stage in the biome's evolution they develop. For most of our vegetables and perennial plants, a ratio of around 1:1 or 1:2 is thought to be optimum, while fruit and timber trees benefit from a much higher proportion of fungi, reflecting their place in evolutionary history.

Climatic factors, such as temperature and rainfall, are crucial to soil development and soil life. Warmer temperatures speed up the chemical reactions that help to break down the bedrock and, along with increased rain, help plants to grow, which increases the biology of the newly forming soil. In farming systems, though, how we manage our soils also has an impact on their fungal to bacterial ratio.[8] The things that are particularly damaging to fungi are disturbance, such as ploughing or other cultivations, high fertilisation rates and a gradual reduction in soil carbon (often connected to the first two factors). Using fungicides, as the name suggests, can also reduce soil fungi populations. There is an increasing recognition of the importance of healthy fungal populations in soil, but finding ways to shift the ratio is not always straightforward, particularly in intensive production systems.

Cutting out fungicides helps, but many farmers are locked into a system that relies on them for high yields. These systems are usually also dependent on nitrogen fertiliser. It is well known that high levels of nitrogen fertilisation (either artificial or from natural materials) encourage quick growth. This rapid expansion can make plants more susceptible to fungal infection. Reducing fertiliser often leads to a reduced need for fungicides and insecticides. Interestingly, this effect seen on individual plants seems to be replicated when you zoom out to a wider system-scale view. One study found that increasing fertiliser application led to a systems failure that increases the risk of pathogen infection across the plant community.[9] It appears that those species that are best at fighting fungal diseases are those that do less well in higher-nitrogen conditions. As you increase nitrogen levels, those that suit the higher levels outcompete those that are more resistant to pathogens, leading to a shift in relative population.

Another way to build fungal abundance is to reduce physical disturbance of the soil. This is difficult for many farming systems to achieve.

There has been a huge increase in the number of farmers adopting no-till and minimum-till approaches, but many of these still rely on the use of a herbicide to kill the previous cover crop or weeds. Glyphosate is the most widely used of these herbicides and has been shown to influence the development of mycorrhizal fungi in soil.[10] This negative effect may well be outweighed by the benefit of leaving the ground unploughed. In cold or very dry climates there are mechanical methods of destroying cover crops that eliminate the need for herbicides, but in the warm, damp climate of the UK this is far trickier. A group of farmers have been trialling this through the Innovative Farmers Programme, but more work and trials are still needed.[11]

Organic farming has a fertility building phase as part of most rotations. This is normally a clover or grass ley, sown and left for a year or two during which time it is grazed or mown. This ley allows the soil to recover, soil carbon to increase, and microbe and fungal populations to revive. Farms with longer rotations, often those certified as organic, have been shown to have higher soil organic matter levels in their soil.[12] What we would really like, though, is to be able not just to maintain those organic matter and fungal levels but to enhance them.

Gardeners and small-scale growers are increasingly exploring 'no-dig' systems. As well as long-time proponents such as Charles Dowding in the UK, there are numerous new celebrity growers marketing highly intensive no-dig systems as a way to make a living by producing vegetables on a small area of land. The economics usually only work in our current food system by concentrating on high-value crops like salad leaf and micro vegetables. It is very hard to compete on price with larger producers on bulk commodity crops like carrots and potatoes. In these high-output systems, there is often not a resting phase in the rotation; instead they rely on the reduced cultivations and added compost to increase soil health. This approach also helps to shift toward a higher fungal abundance and diversity.

As we can see there are ways to build your soil fungi, but most require some significant systems change; that can be daunting, or perhaps not possible for some farmers for certain crops or land. This is where woodchip comes in. Woodchip is fungi food, and by its very nature encourages an increase in their number and diversity. Woodchip gives a lasting effect to the soil biology, since it is made up of complex lignins and takes time to break down.[13] While green compost and straw-based organic amendments may give a quicker response, as they are more readily broken down, woodchip would appear to provide a longer-term shift in fungal to bacterial ratios. Soils with a higher fungal ratio have also been shown to have a greater capacity to store carbon.[14]

Ramial Chipped Wood

Ramial chipped wood (RCW), translated from the French *bois raméal fragmenté*, is the name given to woodchip that is derived from the smaller twigs, branches and stems of trees and hedges. The standard definition is that it is a chip from material less than 7 centimetres (about 3 inches) in diameter. What differentiates this type of chip from other forms is its high cambium to cellulose ratio. This means it has a softer structure and a greater nitrogen (and some other key nutrients) percentage than older wood and breaks down more readily. The potential to exploit these properties for the benefits of soil health first started to be explored in the 1970s in Quebec by Lionel Lachance, Alban Lapointe and Edgar Guay. In the early 1980s, Gilles Lemieux joined them, and it was then that the term *bois raméal fragmenté* was coined. A decade of research provided good evidence of the benefit of using RCW to build soils, but the topic was not then further researched and, surprisingly, has not found wider resonance until recently. The most recent study into RCW – a project called Woodchip for Fertile Soils by the Organic Research Centre in the UK – has just concluded, providing some further evidence for the practicality of using this valuable material on farms.[15] In this section, we will look at what evidence there is for the benefits of RCW, and the practicalities of using it.

Lemieux observed[16] forest soils and drew the following conclusion:

The most active elements [of forests] are located in the ramial parts, which produce buds and leaves and, in numerous cases, fruits. By returning into the soil these 'ramial wood chips' rich in energy through biotransformation, they can renew, rehabilitate and, most important, reconstruct the degraded soils. In these soils, the fundamental mechanisms are found and can rebuild a fertile soil where no other modern technology has yet succeeded.

While conceding that they did not fully understand all the mechanisms at work, Lemieux was clear of the benefits seen by using RCW because it is 'made from tree parts, branches, twigs and leaves rich in nutrients, sugar, protein, cellulose, and lignin, which all play a precise and specific role in the formation and maintenance of fertile soils.' In fact, these smaller branches contain up to 75 per cent of this good stuff.[17] Soils with RCW amendments showed several improvements. They made better use of available water; Lemieux describes the soil organisms that the RCW supports as actively managing the water supply, meaning that the plants cope better during drought periods. As well as significant yield increases, RCW appeared to moderate pH, both lowering the pH of alkaline soils and increasing that of acidic ones. They also saw reduced pathogens in crops growing in soil with RCW additions.

How and Why Is RCW Better Than Compost?

Compost is, of course, also great for soil health, as long experience and study show. However, the nutrients in compost tend to be more readily available than those in RCW. This leads to a quicker release and take-up of nutrients, but they also 'run out more quickly'. Composts, and particularly those from high nitrogen sources like manure, also tend to favour bacteria over fungi; this leads to the unpleasant smell of poorly managed compost heaps. The other big advantage of RCW is that it takes no time to process; you can chip and immediately spread. This cuts down the cost and time associated with producing a high-quality compost. Like the engineer Jean Pain in France, the Quebec team sought to take advantage of a material not often used. Though they did recognise that, as with any technique, there are risks if scaled up too quickly. For instance, if you chip all the small wood there may be nothing left for animals to eat. If you are planning to harvest RCW in a silvopastoral system, there may well be a balance to be struck between the different needs of the various enterprises. If you cut all the small wood for chipping, will your animals have enough browse?

Impact of RCW on Soil Biology

Though there are chemicals present in woodchip that can benefit soil health, the power of using most organic materials as soil amendments is their effect on soil biology. This is true for RCW. Lemieux describes the particular benefit of chipping wood as akin to us chopping and chewing our food before eating it to make it easier to digest.[18] In a natural forest, this process happens more slowly as whole branches and trees fall, to be consumed and broken down over time by insects and then fungi. Our agricultural soils tend to have lower levels of biological activity and so 'pre-chewing' this wonderful soil food before adding it helps a quick ingestion and hence quicker benefits.

Unsurprisingly, adding RCW has been shown to benefit soil fungal populations.[19] In both Canada and the UK, increases in earthworm populations were seen. The Organic Research Centre's Woodchip for Fertile Soils (WOOFS) trial also recorded a significantly higher number of bacteria, even compared to the other treatments of green compost and composted woodchip. If we were not also seeing fungal increases this bacterial blooming might be seen as a problem; however, as part of a general increase in soil life, the researchers interpreted it as 'an indicator of abundance of food for predators, nutrient capacity and general diversity of the bacterial population and the health of the soil, suggesting some positive effects of the RCW over the compost or control treatments.'

Impact of RCW on Phosphorus

Phosphorus is one of the most crucial nutrients for crop growth, but is not always readily available to plants. It gets locked up in both overly acidic soils and overly alkaline soils. Measures of phosphorus abundance are, as a result, not necessarily measures of phosphorus availability to plants. In systems reliant on artificial fertilisers, farmers overcome this by adding either superphosphates (a soluble form of phosphorus) or substances like rock phosphate that release their phosphorus more slowly. Neither of these are available from sustainable sources in the long term and there have been concerns for some time about the world stocks of these products. They come from a very limited number of places, not to mention the environmental and social impact of their mining and transport.

We know that biologically active soils make what phosphorus there is already present in the soil more available to plants, so at the very least we would expect the higher levels of microbial activity associated with RCW additions to help with phosphorus availability.[20] However, the other crucial actors here are the fungal mycorrhizae, which, when functioning properly, increase plant-available phosphorus.[21] There is an enzyme, called alkaline phosphatase, that plays a special role within the process of making phosphorus accessible to plants. This enzyme is present in very high concentrations in some RCW. *Quercus rubra* (red oak), for instance, is particularly rich in these phosphatases, though it is not fully understood why. In his lectures 'The Hidden World That Feeds Us: The Living Soil', Lemieux talks of unpublished work showing an increase in soil phosphatase levels when they are treated with RCW. It is possible, therefore, that we could harness this ability of RCW to help solve the phosphorus problems in some arable farming, by making full use of the extensive, but currently unavailable, soil reservoirs without having to resort to external inputs. A note of caution, however: over-use of organic materials, and particularly woody materials, can cause excessive levels of phosphorus (and potassium) that may lock up other nutrients. As ever, looking for balance in our systems is the key.

Which Species of RCW?

Using RCW from conifer trees is problematic, probably because of the range of polyphenols and other aromatics present in conifer species.[22] 'Climax' species, or those that represent the final phase of forest develop-ment, seem to produce the best RCW, while for most species of conifer it is not recommended to include them at a rate of more than 20 per cent, though the deciduous conifer species larch seems to be an exception. In practice, the types of wood likely to be most used in a farming or garden situation are quick-growing deciduous species that are normally coppiced or trimmed. Most of these are well suited for RCW.

Spreading RCW

The method recommended by Lemieux is to spread with a manure spreader at a rate of 150 cubic metres per hectare, which equates to a depth of about 15 millimetres (just over ½ inch). Unlike some others, he also suggested incorporating the chip to a depth of 10 centimetres (4 inches) by harrowing. After this initial input, he advocates adding half that volume every three years, to ensure the soil fungi and other organisms are fed and soil health is maintained. Before the current fashion for describing oneself as a regenerative farmer, Lemieux was using the term 'soil upgrader' to describe someone who was not merely looking after their soil but actively building it.

On one of the WOOFS trial sites, they tried adding 40 cubic metres (50 cubic yards) per hectare in the first year, and then doubled the rate in the second year of the trial on half the plot. This led to a significant increase in organic matter and phosphorus in the higher application plot. This might suggest that the biology was stimulated in the first year to a point that it was able to process the larger application in year two.

There are so many variables to consider when looking at applications – such as soil type and health, rainfall and the type of chip being used – that giving a definitive recommended rate to spread is tricky. The team at the Organic Research Centre settled on a rate of 60 cubic metres (80 cubic yards) per hectare as their standard rate for the calculations, though it is likely that if you are trying to regenerate degraded soils you might start with a higher rate, reducing both rate and frequency of spreading in subsequent years. As with any new strategy, though, it is always better to do a small trial first. Putting this relatively high rate on in one go to an arable field that you need the yield/income from is a risk. Perhaps try that rate out on a strip in year one, then if you see no problems, apply at that rate in the field in the following year. Alternatively, you might go for a more conservative 30 cubic metres (40 cubic yards) per hectare in year one and 30 cubic metres per hectare the following year.

Timing

Even with RCW, there is a small risk of nitrogen lock-up. Timing your application can help eliminate this risk. Though spreading any amendment must fit around other farming and gardening jobs, in the WOOFS trial all the farmers spread during the winter. This gives the wood time to break down a little before plants start growing. One farmer added some mineral nitrogen at the same time to ensure there were no shortages, while the organic farmers in the trial added the RCW to a growing legume ley. The advantage of this latter strategy is that the soil will be biologically active and able to more immediately start the decomposition process. In addition, once the ley is incorporated for subsequent crops, there will be a further release of nitrogen that organisms can make use of to break down the wood more quickly.

Producing Your Own RCW

A main selling point of RCW is that it is a material not currently highly valued and likely to be available either on your own holding or from other local sources. All the farms in the WOOFS trial had access to plenty of RCW, and the team were able to do some calculations regarding how many trees you would need to produce your own woodchip. In the absence of specialist hedge-trimming equipment that collects the trimmings, the easiest way to produce RCW from existing tree stocks is to coppice hedgerows. In Table 5.2, you can see an estimation of resource needed to produce enough chip to treat 1 hectare (2½ acres) of land each year with 60 cubic metres (80 cubic yards) of RCW.

Further observations from those farmers involved in the trials suggest that if you plant with the intention of producing RCW, single-species rows are the most practical and efficient approach. Willow seemed to be the most popular for its rapid growth rate and ease of chipping. Though there were differences in the chemical analysis of RCW from various tree species, this difference did not appear to have any noticeable effect once the chip was spread. It seems, therefore, that you can use anything you have available, or plant what is most suitable for your site, and convenient to manage and harvest.

The team also did some cost analysis of buying in or producing your own RCW from various sources and compared to alternative inputs of green compost and woodchip compost. Costs ranged from about £650 ($900) per hectare up to nearly £2,500 ($3,500) per hectare. The highest figure, though, was for producing RCW from existing hedgerows, and so the cost of alternative hedge management strategies can be deducted from this. The variation in these costs came down mostly to scale and available machinery, though the system from which you are harvesting will also have an impact. For instance, harvesting single-species straight-row coppice will be cheaper than managing a mixed hedge with large trees in it that interrupt workflow.

The Potential for RCW

There is still much to learn about using RCW. There is no doubt it has massive potential to harness a material with little current value to significantly boost soil health, increase yields and help to mitigate climate

Table 5.2. Estimation of Resource Needed to Produce Enough Chip to Treat 1 Hectare (2½ Acres) of Land Each Year with 60 Cubic Metres (80 Cubic Yards) of RCW

System	Coppice Rotation	Average (Range) Volume of RCW (Cubic Metres)	Total for 60 Cubic Metres RCW	Total Length / Trees in Coppice Rotation for 60 Cubic Metres RCW Annually
100 metres mixed species hedge (100–200 trees)	10 years	12.75 (8–17.5)	470 m	4.7 km
100 metres hazel short rotation coppice agroforestry row (133 trees)	5 years	11 (8–14)	546 m	2.7 km
100 metres willow short rotation coppice agroforestry row (165 trees)	2 years	5 (4–6)	1,200 m	2.4 km
Brash per tree cut for fuel logs (willow)*	7 years	1.14	52 trees	370 trees

* Plus around 0.5 cubic metre of logs per tree.

Assumptions:

1) Volumes based on actual figures from RCW trial sites and established systems.
2) Volumes based on actual figures from RCW trial sites and established systems.
3) Hedges are allowed to grow upwards and coppiced at approximately 7 m height.
4) Brash chipping assumes that all material with a diameter greater than 7 cm is taken for logs.

change. Perhaps the quickest win would be in helping to shift hedge-row management from annual flailing to a longer rotational coppicing strategy. Planting less productive patches on farms with short rotation coppice to feed more fertile areas could also be a game changer. In the current debate about reducing meat in diets and therefore gradually reducing livestock numbers, you could also see how bringing areas of coppice onto land that is not suitable for arable and horticultural crops could be used to support higher output from the better growing land.

It does seem that farms with longer rotations and fertility leys are more suited to using RCW, as it can be added during the ley phase with less risk to following crops, though adding small additions of mineral nitrogen, slurry or digestate can reduce potential disruption to fertility within a more intensive farming system.

Some questions remain. We do not fully understand the complex relationships between RCW, the soil and the organisms that break it down. There were some anomalies in the WOOFS trial, for instance, where some crops showed a higher yield and others a lower from the same treatment. Observations, such as the reduced slug number in the potato crop with RCW, indicate that there are likely to be some very specific interactions that might offer benefits, or not, depending on crops or soil types. They even observed reduced growth of some brassicas early in the season three years after the RCW application, though this did not seem to affect final yields.

The WOOFS team summed up their study by saying that: 'RCW is not a panacea, but has the potential to offer some significant benefits in terms of carbon capture and storage, overall soil and crop health as well as helping farms move toward self sufficiency in inputs and closed system farming.'

Worms

There's more to soil life than earthworms, but they are still a good indicator of soil health, and an easy one for farmers and gardeners to look at without needing microscopes or soil testing in the lab. Earthworms were famously christened nature's plough by Darwin in his amazing and very popular 1881 book in which he observes and details worm populations, burrow dimensions and the quantity of organic material that passes through their digestive systems.[23] But what evidence is there that using woodchip will build worm populations, and are there things we can specifically do to help support our worms?

Earthworms consume and break down organic matter. The bacteria in their gut are well tailored to breaking down plant cellulose, including that in woodchips. They also play a role in moving fungal spores around the soil along with the digested material they excrete. Worms do not have teeth; they effectively swallow organic material, coating it in saliva and pulling it down into their gut. Size and shape of the ingested particle have an impact on their ability to eat it. Darwin observed that worms specifically pulled pine needles down into their burrows the right way round to avoid them getting caught. The needle itself is very long, but thin, so easily swallowed. Fat woodchips present more of a problem for them. For this reason, you will often find worm composting sites advising against using woodchips in worm-based composting systems. While this is undoubtedly true if your main aim is a quick turnaround, high worm population composting operation, there is no reason to think that woodchip in compost more widely will harm worms.

If you have ever observed the gradual decomposition of a fresh pile of woodchip, you will have noticed that it takes a while before the worms move in. If that pile is on soil, they don't take long to start exploring the new pile for small bits of leaves or wood that they can manage. Even for piles on concrete, worms seem to find their way into the heap once the fungi start rotting the wood fibres down to a point that they are edible for worms.

A study from the 1950s looked at worm populations in plots with hardwood woodchip amendments combined with differing fertiliser additions. The findings were clear: 'hardwood chips increased the number of earthworms by a factor of 3.'[24] They also proposed that, 'the earthworm counts appear to provide indices of the general biological activity in the soils. Their life cycles may have been important factors in the differential conservation and build up of soil nitrogen shown by soil tests.' The study found that while the differing fertiliser treatments did have an impact on vegetable yields in a given year, at the end of the study the soil available nitrogen was more dependent on woodchip than on the fertiliser.

More recently, the WOOFS trial looked at worm populations in relation to amendments of ramial and composted woodchip.[25] At one of the trial sites, Tolhurst Organic, the researchers found that, 'across all trials and years, the total worm number was significantly higher in RCW and compost plots than the control plots with significantly

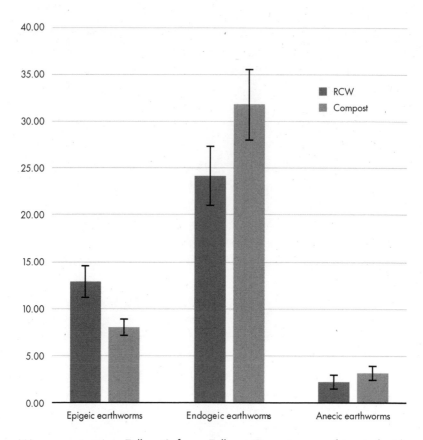

Worm count at Iain Tolhurst's farm, Tolhurst Organic, on soil treated with RCW or composted woodchip treatments. *Source:* S. Westaway: WOOFS Technical Guide 2 2020 Organic Research Centre.

more epigeic [surface-dwelling] and anecic [deep-burrowing] worm ecotypes also seen in the RCW and compost plots compared to the control'. Total numbers of worms in both the woodchip plots were also impressive: 793 worms per square metre, or nearly 8 million per hectare. This is somewhat more than the figure that Darwin quotes of 133,000 per hectare. Worm populations in the other two farms in the WOOFS trial, which were both arable systems, were around half of that; however, that still worked out as about 3.5 million worms per hectare.

Species diversity was also higher at Tolhurst Organic (eleven species) compared to the arable farms (eight and six). The ramial woodchip seemed to encourage the surface-living epigeic worms, while composted material resulted in higher populations of endogeic (burrowing) species.

The conclusion, both from personal observations and from these studies, is clear: using woodchip will increase the worm population in your soil. However, more than that, it seems probable that those worms are necessary for the soil to benefit more widely from the additions of woodchip. Whether chip is applied as a mulch or spread more thinly, it is worms that do most of the work bringing it down to the lower levels of the soil and, working closely with fungi and their gut bacteria, breaking it down, making those nutrients accessible to other soil microbes and subsequently your crops. In the long term, those additions of woodchip processed by worms and absorbed into the soil will increase available soil nitrogen.

No-Dig Growing

Where once double digging and deep ploughing were the norm, there has been a move in both farming and gardening toward reduced soil cultivations. We are all now more aware that a minimum-intervention approach brings significant soil health benefits. In arable farming systems, moving to a minimum- or no-till system relies on the judicious use of cover crops. In more intensive horticultural and garden situations no-dig methods using compost, including composted woodchip, are a great option. Charles Dowding is a leading proponent in the UK, and I would highly recommend his books and YouTube channel for a more detailed demonstration. Here I will give you a brief overview and look at how best to use woodchip.

Basic Principles of No-Dig

As the name suggests, the main principle of no-dig is that you sell your spade and leave the soil alone. Mixing different layers of soil disturbs soil fungi and microbes. If you turn the soil over, you are effectively moving those organisms from the part of the soil profile where they are happily at home, to a place (lower or higher) where they might not want to be. Time spent digging is time and energy spent decreasing your soil's health.

Woodchip on no-dig beds. Photo courtesy of The Wild Croft in Highland, Scotland.

There are also a huge number of weed seeds in the soil. When you disturb the soil you are bringing more of those seeds to the surface, both allowing and encouraging them to germinate and creating more work for yourself. Though of course some weed seeds will blow in from elsewhere, minimal soil disruption reduces your weed burden considerably if you can prevent those weeds that do grow from flowering and seeding. The old adage of 'one year's weeds, seven years' seeds' holds true.

There are times when you still might need to resort to digging; such as removing large perennial weeds, or dealing with historic compaction, but, even in these instances, the aim is not to turn the soil, only to break through its layers. As your no-dig system matures, you should need to do less and less.

The key to a successful no-dig garden is the prudent use of mulch to keep the weeds under control and feed the soil. If you are transforming a lawn or weedy patch to no-dig, it is worth putting a thick layer of newspaper or cardboard down in the autumn, adding the mulch on top. By the following spring most of the weeds should have died off and the paper or cardboard will have rotted away. Some perennial weeds are likely to have survived, but these are easier to remove from the organic rich surface. If you have an area with a lot of bad perennial weeds, it can be worth the effort to dig them out before you start. In the early stages of developing your no-dig garden you will need to add a thick layer of mulch to control the weeds, but as the system matures a thinner layer should be sufficient to maintain the soil health.

Most organic materials can be used to create a no-dig system; green compost, various manures and mushroom compost are all good, and, of course, composted woodchip. I would not recommend mulching vegetable beds with uncomposted woodchip (unless it is ramial chipped wood) as there is a risk of nitrogen lock-up. You could mix the woodchip with a high-nitrogen material like manure, but it is safer to compost the chip for a year or two first.

The no-dig technique relies on soil organisms to break down and incorporate the organic matter into the soil, which is indeed what happens in nature where leaves and other plant material always fall on the surface. The top layer of the soil becomes very rich in organic matter, and the almost constant presence of a mulch brings all the benefits that we explore in chapter 6. As well as significantly reducing weeding, no-dig systems tend to remain workable and dry on the surface. This often allows you to work on the soil for more days in a year and minimises damage when harvesting in wet weather.

Aminopyralid

In both the US and the UK there have been reports of contamination of manures with the weed killer aminopyralid. It is a persistent chemical used to kill broadleaved weeds in grassland. However, it can survive through the animal's gut and persist in the resulting manure. This, when spread on soil, affects the health and growth of subsequent vegetable crops. One advantage of using a woodchip mulch is that you can reduce the risk from this poison; mixing manure with woodchip and turning will also speed up the breakdown of the chemical if it is present in the manure. Mixing into the soil also helps to break it down. One disadvantage of no-dig systems is that they may suffer more from aminopyralid contamination as the material is not incorporated into the soil but sits on the top where it takes longer to disintegrate.

Other Potential Issues with Using Woodchip

Charles Dowding found that using uncomposted woodchip mulch, even on the paths, can encourage woodlice and other creatures, which can damage some crops like spinach or cucumber. Slugs are also reported to be a problem using woodchip, though I have to say I have not found this; if anything, on my allotment the woodchip seemed to reduce slug incidence.

Due to the high phosphate and potassium content of wood, there is a risk from continuous use of increasing levels of these nutrients to a point where they begin to lock up other nutrients. In a normal rotational system, you would not add woodchip every year, so this problem is unlikely to arise, but in a no-dig system you are reliant on annual mulching. Proponents of no-dig systems claim to see no detrimental effects from this, and it is possible that if you are harvesting large amounts of crop, you are taking enough nutrients out of the system to prevent this imbalance forming. It is worth doing occasional soil tests to ensure that this problem is not building up. Some growers also do nutrient balance calculations, estimating nutrients leaving the soil as crop compared to those being brought in.

No-dig is a great method for reducing labour and improving soil quality. In temperate climates, it improves drainage and controls weeds, while in poorer soils and in drier conditions it can be used to make better use of water and reduce the risk of erosion.[26]

Hügelkultur

The term comes from the German meaning mound or hill culture and describes a specific technique of growing on a raised bed. Unlike more traditional raised beds which are made simply by mounding up the soil, often with a retaining wall of wood or other material, *hügelkultur* builds up layers of woody material first before adding a layer of compost and soil on top. It is thought that it was first developed following observations of how well plants grow in soil with lots of woody additions. In other words, those pioneers were attempting to recreate forest conditions.

There are many variations on how exactly to build a *hügelkultur*, but the basic concept is this. First, dig out a trench, piling up the soil carefully for reuse. In the trench, you lay a mix of branches and logs; this will leave lots of gaps, and this is where our woodchip comes in. Pile woodchip, leafmould, compost or any other similar material onto the bottom layers of wood. This layer could be fresh or composted woodchip. Once your mound is the desired shape, add a layer of well-rotted woodchip or compost before finally topping off with the soil you dug from the trench. Plants can be sown or planted into the soil, making use of the nutrients from the gradually decomposing organic matter underneath.

The beds are normally about 1 metre (3 feet) in height when constructed, though, of course, as all the organic material rots down the heap will gradually become smaller. Depending on the size of logs used, the beds tend to rot down totally after five years or so, at which point you start again. They share benefits of other forms of raised beds – good drainage on wetter land, for instance – but because of the levels of organic matter contained in the mound, they are particularly good at holding lots of moisture and so are especially recommended for drier climates with sporadic rainfall.

There is not much scientific evidence on this technique of growing, though anecdotal and unpublished research projects do suggest benefits. There is a risk of having too much available nutrient within the mound, especially if too much woodchip or compost is included in the construction.[27] At worst, this could cause leaching and contamination; using uncomposted woodchip could reduce this risk as there would be a delay before the nutrients were released. For some crops, these high nutrient levels might cause excessive growth and weaker plants.

Despite the high carbon content of the woodchips included in the mound they do not seem to cause nitrogen lock-up, perhaps because the top layers have plenty of nutrients, and by the time that might run out the lower levels have decomposed enough to replenish.

Restoring Damaged Soil

With the potential for woodchip to boost soil health, hold onto water and promote plant growth, it is a small step to look at how to effectively harness that potential for rescuing degraded and damaged soils. There are numerous examples of how this has been done, and we'll look at a few of them here. In some of the studies woodchip and biochar are used either comparatively or in combination, and there is certainly potential for combining the shorter-term benefits of woodchip with the longer-lasting properties of biochar. Both the physical and biological properties of woodchip are used in soil remediation.

Bioremediation

We have already seen the potential for woodchip and the fungi that decompose them to absorb potentially polluting nitrates, but there is some evidence that it could be used more widely to help deal with other manmade pollutants, 'such as chlorinated and non chlorinated hydrocarbons, wood preserving chemicals, solvents, heavy metals, pesticides, petroleum products, and explosives.'[28] There is an even stronger body of evidence on the potential of biochar for this purpose, but creating biochar is more costly and in most cases some of the energy is lost during the production process. Woodchip is cheaper and easier to produce, so it is worth looking at those situations that could use it.

For contaminants that would eventually break down anyway, such as oil and diesel spills, woodchip appears to have the potential to significantly increase the speed at which the contaminants disintegrate. One study showed that adding woodchip at 50 per cent or more boosted the levels of microorganisms capable of breaking down crude oil.[29] Another project looked at compost, fungi and growing willow to remove diesel contaminants from the soil.[30] This method performed better than the industry standard chemical treatments. The woodchip effectively feeds those organisms that can degrade the oil pollutant, multiplying their numbers and so increasing the speed at which they can deal with the problem.

There has been some work looking at using woodchips to help reduce the salinity of soils. They could have a role in preventing damage, for instance in protecting roadside plants from salt spread for de-icing, or in reducing the salination effect of animal manures.[31] Woodchip could even help with reclaimed soils that have excess salt levels. An Australian study found that adding woodchip as part of a combined strategy helped to improve soil structure, enabling increased leaching of salts from the soil surface.[32]

In practical terms, in a farm or garden situation we can look to use woodchip if we have an accidental spill of oil or diesel; or, if you are taking on a site with some history of contamination, using composted woodchip may be a way to reduce the risk associated with this. Contaminated soil fed to rats with and without compost additions showed that the compost held onto the contaminants more tightly than untreated soils, with rats showing no toxic side effects of the compost-treated soil compared to some effects with untreated.[33]

Woodchip bioreactor in Monmouth, Illinois, US. Photo courtesy of Laura Christianson.

In areas that have a history of surface mining, for instance for zinc or lead, this a treatment worth considering.

Woodchip Denitrifying Bioreactors

In the context of soil remediation this type of bioreactor is different from the Johnson-Su reactor described earlier. Here, the term is used to describe a technique of using woodchip to absorb nutrients (and particularly the soluble nitrates and phosphates) lost from farmland. Draining farmland is an important way of making it more productive and suited to cropping and grazing. The Romans used field drains, and in the eighteenth century new designs using bricks and tiles were extensively laid down. Nowadays, land drains are mostly plastic perforated pipe, but the principle remains the same; provide a channel to remove water quickly from wetter fields to prevent anaerobic conditions. One significant consequence, though, is that if there are nutrients in the soil water these will be washed away through the drains and into the rivers. If you force this draining water through a woodchip bioreactor, you can prevent most of that soluble nutrient reaching the river and return it to the land by furnishing a carbon and energy supply to denitrifying microbes.[34] The water needs at least twenty-four hours in the woodchip to effectively take most of the nitrates out, and the woodchip needs to be regularly replaced to stay effective, usually after about four years. On a farm scale you can construct a simple lined bed of woodchip, which you can then cover back up with soil.

Compacted Soils

Compaction is a real issue in many cultivated soils and is often solved by subsoiling, usually involving a deep tine dragged through the soil to break up the compaction. While this does fix the problem temporarily, without a more permanent change to managing the soil, the symptoms will inevitably reoccur. Along with other approaches such as lengthening rotations, growing deep-rooted green manure crops and reducing tillage, adding organic matter will help to reduce compaction. Woodchip has been shown to be effective in restoring wetlands, for example, where other methods might not work, by reducing compaction.[35]

Similarly, in urban soils where there is heavy foot traffic, and little in the way of soil management, compaction is a big challenge. It has the biggest impact on urban trees, which often already have limited soil and water available. In many cities the ever-growing financial pressure of managing street trees has led contractors to pave ever closer to the trunk. However, even where there is still some soil around the tree, there are few options available for effective soil management, given the very tight spaces and lack of available suitable machinery. A study at Morton Arboretum in Illinois concluded that, 'compost top dressings and wood chip mulches should be used as soil management techniques for trees growing in compacted urban soils,' and that, 'it is reasonable to expect that combining wood chips and composts may have even greater benefit for improving soil quality for urban trees.'[36]

Erosion

Many of the studies done on using woodchip for erosion control have looked at extreme situations, such as post forest fires or after large areas of trees have been clear-felled. While this is not a replica for most farm and garden scenarios, we can use the knowledge and insight gained from this research to help on smaller-scale erosion challenges, like growing vegetables on sloping land or protecting cultivated arable land.

When we talk about degraded soils, we normally mean that they have lost structure and organic matter. Usually, this is associated with heavy and continuous cropping. In a reflection of how a soil builds from nothing with the gradual undisturbed addition of organic material and subsequent increase in life, if we cultivate and grow crops without adding organic matter, we will gradually reduce that soil back to its mineral content and structure. Without the carbon, air and water held within the soil that are needed to support life, eventually it can become a desert.

Changes in agricultural practice tend to have a greater effect on soils in drier climates where quick recolonisation of weeds over bare soil is less likely. On slopes, this can quickly lead to soil loss and further challenges of establishing plant cover. Using a woodchip mulch significantly reduces that erosion and helps to build soil structure and moisture-holding capacity of the soil. In a study in northeast Spain looking at gypseous and calcareous soil post forest fires, a thin mulch, of between 0.5 and 2 centimetres (⅕ and ¾ inch), was shown to considerably reduce runoff both of water and sediment.[37] They found that you needed to mulch at least 60 per cent of the site and recommended mixing seeds of a suitably adapted plant species into the woodchip. In a farming situation, you could potentially mulch with a thin layer of RCW or composted woodchip on land that was at risk of wind or water erosion. This could be done post-harvest if weather conditions allowed, though it would be tricky in wetter climates after a late-harvested crop. Depending on the following planting, a quick-growing cover crop seed could be added into the woodchip.

CHAPTER SIX

Mulches

U sing woodchip as mulch is probably its best-known use. Many tonnes of chip are spread around playgrounds, herbaceous borders and garden paths every year. We'll look first at the well-known properties of weed control and water retention that mulches are generally used for, but there is increasing evidence that adding woodchip can bring longer-term benefits for soil and plant health, as well as even the potential to stimulate protection against diseases. Though more studies are needed, there is some great work done by David Granatstein in the US looking at the commercial implications of using organic and woodchip mulches, while Glynn Percival in the UK has been exploring the role of single-species mulches; for instance, how the salicylic acid in willow chip might affect disease immune responses in trees.

Water

One of our main challenges as growers is how to manage water for our crops. We'd like to hold onto and find water when there is too little, while in times of excess the challenge is getting rid of it or trying to collect it for later use. With the increasingly erratic weather patterns that climate change appears to be throwing at us, this challenge will only increase. I created an unintentional experiment at Eastbrook Farm when we dumped a load of woodchip near some newly planted trees, meaning to spread it out in due course as mulch around them. As is often the case, other jobs were more important and we never moved the piles, which were around 3 feet high. This was during the very hot, dry year we had in the UK in 2018. The resulting tree growth after two years under the mulch was striking compared to those that had no mulch, as can be seen in the photo on page 7 of the photo insert. After two years, the mulched willows had grown to about 8 feet tall, while those without had barely reached out of their spiral guards. My guess is that the main reason for this difference is the mulch prevented the ground from drying out. In this field, the soil is heavy and deep, so there was still some moisture in the deeper layers, even at the height of the summer heat. Perhaps the chip also kept the soil warm in the early spring, and then cool in the subsequent summer heatwave? We did suffer some cracking of the soil in the unmulched soil, which no doubt also contributed to the suffering of those trees without a chip protection. We will explore the more general role that increasing soil organic matter can play in this, and how using woodchip more widely might help that. For this section, though, I want to explore how mulches can help us to manage water supply and demand.

Holding It In

Organic mulches can keep water in the soil and absorb it, holding on to and then releasing moisture.[1] This is usually advantageous, slowing down the passage of rain to the soil. But if you are planning to irrigate your plants once you have mulched, remember that some of that irrigation water will soak into the mulch and then evaporate back into the atmosphere. For plants that need watering, especially at establishment, you can mitigate this risk with one or more of the following methods:

- Soak thoroughly before irrigation.
- Add an irrigation hole, such as a pipe that goes through the mulch to the root zone.
- Lay drip irrigation pipe under the mulch.
- Use a coarse bark that absorbs less water and allows it to run through to the roots.
- Use a thinner layer of mulch to prevent it soaking up too much water.

There is also a risk that the underlayer of broken-down chip will form a barrier which the water can't penetrate.[2] In this scenario, giving the mulch a quick rake to break up that underlayer will help with infiltration. When insufficient water gets to the soil the result can be a moist mulch and a dry soil, which risks encouraging the plant roots to stay at the surface rather than growing downward to find moisture. Then, when the mulch dries out, the roots have no access to water and the plants will suffer. In heavy soils it is unlikely that the lower levels will dry out, but in light soils in dry climates this is a significant risk.

One study looked at using a wood fibre mulch.[3] These woody fragments were longer than a traditional chip material would be, with an adhesive added in to allow the mulch to form a more solid barrier and prevent blowing away. The significant reduction in evaporation they observed compared to bare soil led to increased soil moisture and a resulting increase in yields and quality. This result is backed up by another study where a 15 centimetre (6 inch) mulch of hardwood chip led to a 26 per cent increase in soil moisture.[4]

Reducing Irrigation Requirements

There is a large body of evidence that organic mulches will reduce evaporation from the soil and reduce the need to irrigate. Specific research on woodchip is less easy to come by, despite its common use as a landscaping material. One study in orchards in the Pacific Northwest found that woodchip mulch spread at 10 centimetre (4 inch) depth led to a 20 to 25 per cent savings in irrigation water.[5] Another trial with blueberries provided different watering regimes and mulches and found that, 'differences between irrigation treatments were minimized with mulch.'[6] They also found that mulches increase the yield when drip irrigation was used, but not with spray irrigation. With spray, the mulch is likely to absorb much of the water, while drip irrigation will trickle through the mulch to the soil where it can be used by the plant.

This ability to reduce the need to irrigate can be crucial for establishing trees or other plants in areas where watering is not feasible. We experienced exactly this when planting trees at Eastbrook Farm in the drought year of 2018. Those trees that had a mulch soon after planting survived, despite the lack of rain. The trees that remained unmulched or received only a very thin layer of mulch had a much higher mortality rate, with nearly 50 per cent losses in some areas.

Preventing Splash

When large water droplets hit bare soil during heavy rain, they damage the soil structure and significantly increase the risk of disease on plants.

As the rain splashes on the wet soil, droplets of dirty water splatter onto the leaves and bark, taking with them soil-borne pathogens, which can then infect the plant. An absorbing mulch like woodchip will soak up the water. Any splashes that occur will not have been in contact with the soil, so are less likely to cause disease problems. In addition, the mulch should create an environment at the soil surface with more beneficial organisms and fewer pathogenic ones, so that any splash from areas where the mulch is thinner will also have lower concentrations of harmful fungi and bacteria.

A trial by Saunders Brothers in Virginia on boxwood trees showed that a mulch reduced incidents of box blight by up to 97 per cent. Whichever species the mulch was made from, it seemed that adding a thin layer (up to an inch) in spring was effective. They believe there were two things happening to reduce the levels of disease. Firstly, the mulch created a physical barrier between any old infected leaves that were still on the ground and the healthy new leaves on the plant. Secondly, the mulch reduced splash from the infected soil under the plants, preventing reinfection. The risk for box is particularly high. It is a perennial plant and therefore in the same spot every year. This enables disease levels to build, unlike rotational crops where there are opportunities to break disease cycles. Box is also a low-growing bush and close to the ground, particularly when used as a short hedging plant, so any splashing is almost guaranteed to hit the leaves.

Preventing Puddling, Compaction and Capping

There are many factors that can cause or exacerbate soil compaction. Using woodchip as a mulch can help reduce damage, and in some circumstances prevent it. For more serious water problems, you will need to find solutions that might include field drains or planting suitable trees to increase infiltration through the soil. However, for what might be termed 'surface' or 'topsoil' compaction, there is a role for using woodchip. Let's look at a few situations to illustrate.

BARE SOIL SURFACE COMPACTION AND CAPPING

Though it is best to keep a green cover and avoid bare soil, there are times within cropping rotations when it might be essential to cultivate and leave the soil exposed. Heavy rain can damage that soil structure, particularly on clay and silt soils, leading to runoff and puddling. Even a thin layer of mulch will protect the surface and allow more gentle infiltration of rain into the soil surface. This mulch will also allow you to walk on the soil with less risk of impairing its structure. A thick mulch might even allow you to drive small machinery over the soil, though this won't entirely eliminate the risk of compaction if the soil conditions are wet.

PAN LAYERS

Soil pans are usually associated with the use of ploughs or rotavators, especially where the ground is cultivated at the same depth over multiple years. The implements cause smearing at that depth, especially on heavier clay soils, leading to an impenetrable layer that will prevent plant establishment and growth, and impede drainage, resulting in surface flooding. Overstocking livestock, especially in wet conditions, can also cause a pan layer. Traditional remedies for pans are subsoiling or growing deep-rooting green manures like fodder radish or chicory. However, resting the soil with a good layer of organic mulch like woodchip can also help. The increased organic matter encourages worm activity, and the larger anecic worm species that are strong enough to break up soil pans given the right conditions.

GENERAL POOR SOIL STRUCTURE
– REDUCING BULK DENSITY

Many soils will not have a surface or subsoil problem but are simply not in good heart due to intensive management, extreme weather conditions or heavy grazing or long periods of exploitation. There are opportunities to improve these soils with woodchip. The 10 per cent reduction in bulk density observed using 15 centimetre (6 inch) hardwood chip mulch by Scharenbroch and Watson, along with associated increases in soil organic matter, demonstrate this potential.[7] Bulk density is the weight of soil for a particular volume. The higher the bulk density, the less air and organic matter there is. Compacted soils tend to have a higher bulk density. Reducing the bulk density allows better infiltration of water, as well as helping to provide a better environment for soil organisms.

PROTECTING HEAVY TRAFFIC AREAS

There are some parts of our gardens and farms that are always a problem, usually around tracks and gateways. A thick mulch of large chip is a useful way to reduce the damage we do by walking or driving. For pedestrian areas, a 10 centimetres (4 inches) mulch is likely to be enough, but for tractor use I would put at least 20 centimetre (8 inch) down, if not more. This is not going to turn these problem areas into perfect soil, but it should help absorb the mud and provided you keep topping the mulch up it will gradually improve the drainage. Larger-sized chip works well in this situation; it takes longer to break down, so the treatment remains useful for longer.

How Thick to Mulch

This is one of those 'how long is a piece of string' questions. There are a lot of variables. The species and size of chip are important, as is the purpose of the mulch; are you aiming for weed control, moisture retention or disease resistance? For many of us, it may just come down to how much chip we have and how far we want it to go. Depth of mulch may not have that much effect on moisture retention. S.J. van Donk et al. concluded from their study that there was little difference in plant growth between their different treatments, though they did see variations in soil moisture levels.[8] Crop and soil type are likely to also have an impact. Despite all these variables, mulch depth does have a direct impact on weed control. Layers of chip thinner than 5 centimetres (2 inches) are unlikely to have much effect on weed numbers.

Weed Suppressant

It was using woodchip to keep weeds down that first got me interested. I hate seeing gardens full of gravel and paving stones, or fields with herbicide patches or strips everywhere. In areas where bark or chips have been used inappropriately, there will likely be no lasting damage, as they will quickly break down, to the general improvement of the soil. However, woodchip is by no means a panacea for all weeding woes; it has limitations and can even have adverse effects, so in this section I'm going to go through what it can and can't do in our ongoing relationship with the weeds on our land.

Rather like a loved but difficult family relative, you can't get rid of weeds. They contribute to our lives but cause us all a lot of grief. I have become more forgiving of weeds as I get older. I used to aim for total elimination, and if you read a lot of conventional gardening and farming books that is seen as the normal, rational, even moral response, but the harm weeds cause is usually very limited. Certain years favour particular weeds. In a rotational system, you rarely get one weed that dominates for multiple years. There are some weeds, such as chickweed (*Stellaria media*), that will wreak havoc on a carrot or lettuce crop but can be easily tolerated amongst larger perennial plants. Low populations of docks and thistles can benefit the soil by breaking through lower potentially compacted soil levels and improving drainage. As with many things in farming, balance is the goal. The key is perhaps to deliberately embrace diversity of weeds in the hope that this will prevent any one weed getting out of control. We can never eliminate weeds, even if we felt it was a sensible approach. By removing one weed species entirely, we merely leave the field open for another to take advantage. Woodchip for me offers a middle ground; a natural way to reduce the impact that weeds have on our cropping, without either creating an environment entirely hostile to them, or using strategies such as herbicides or plastic mulches that cause other wider environmental problems.

To begin, let's look at how woodchip controls weed growth. There are two main mechanisms at work. Firstly, it smothers weeds, blocking out light and preventing them from growing. The chip also creates an unfavourable surface environment for seeds to germinate and establish themselves.

For the first, the key consideration will be the depth of the woodchip mulch that we apply. Broadly speaking, the deeper the chip, the better the weed control. Since there are constraints on how deep we might want to lay the mulch, such as cost, practicality and potential adverse effects on soil health, it is useful to know what the optimum depth of chip is for effective weed control. There is quite a body of research looking at mulch depths. I've picked out a few to illustrate the general point.

S.J. van Donk et al. trialled three depths of woodchip in their 2011 study, looking at 2.5, 5 and 10 centimetre (1, 2 and 4 inch) mulches compared to no mulch.[9] They found that the 5 and 10 centimetre (2 and 4 inches) gave significantly better weed control than the treatments that had none or just the thin 2.5 centimetre (1 inch) layer.

Another trial using 10 centimetre (4 inch) woodchip mulch amongst a range of other options found similarly good effects. These results are further backed up by a study looking at chipped pine and hardwood mulches of 7.5, 15 and 25 centimetre (3, 6 and 10 inch) depth on tree growth and weeds, which found that, 'Weed density and diversity declined significantly with increasing mulch depth.'[10]

These studies suggest a broad consensus that we need to be looking at depths of at least 10 centimetres (4 inches) for effective weed control, though as we will see there are other factors that can affect how well a mulch works against weeds. Applying 10 centimetres of mulch, particularly as an ongoing annual strategy, may not be the right approach for other reasons, such as the potential for raising potassium and phosphorus levels in the soil to such high levels that it causes other soil nutrients to become unavailable. In most situations, deep mulches are a good starting point to get weeds under control, with a more varied management strategy thereafter.

So, we have seen how woodchip mulch can act as a physical barrier to weeds. Hard, slow-rotting species like hornbeam or oak will take longer to break down and therefore provide weed control for longer than a soft, quick-growing species like willow or poplar. In addition, the physical nature of the material has an impact. A coarse fresh chip delivers an inhospitable environment for a seed to germinate. There is a lot of air between the large chips. When a seed lands on the mulch, there is nowhere for its roots to get a hold and the chip is likely to dry out as soon as there is a couple of days of sun, so any new seedling will quickly die. Older chip, or that with a smaller particle size, will perform this function less well, though it might have other advantages.

Mulches with a chip size of more than 4 centimetres (1½ inches) become less effective as a mulch. Although they are good for preventing weed germination, they let so much light through their large gaps between the chips that weeds can grow through the mulch more easily. If you have a material with a larger chip size, you will need to apply it more thickly to get the same weed control benefit. Conversely, a finer chip can be laid more thinly for weed control where as little as

5 centimetres (2 inches) thick might be sufficient. It sometimes forms a crust that prevents weeds underneath from getting through. However, a very fine chip is less effective at preventing new weeds from establishing on the surface. As the mulch rots, it ceases to be a barrier to weeds and becomes a perfect propagation medium for them, potentially encouraging a thick crop of lush green weed cover. Fine chip also composts down more quickly. This makes it less useful as a long-term mulch. Whatever the size of chip, get in quickly with a hoe, rake or harrow when the weed seedlings are still very small to knock them out before they get through to the soil underneath when they will become much harder to dislodge. David Granatstein has seen some promising results by flaming on top of the mulch and running a rolling tine cultivator (Wonder Weeder) and says, 'I have wanted to invent some sort of device, almost like a stiff wire or blade, that could be pulled through the mulch just under the surface and uproot weeds.' This backs up my experience that applying the mulch is not the final act of management, but can be more effective if you continue to actively maintain the surface of the material.

Chemical Effect on Germination

Beyond the physical effects at play, there is also some complex chemistry going on in a woodchip mulch. We have looked at the physical and biological ability of individual wood species to control pests and disease in soils and crops, but we can also exploit their chemical makeup to control weeds. Although allelopathy is usually thought of as a negative property, allelopathic chemicals can produce a beneficial or adverse effect on their subject. For instance, one study found that certain allelopathic extracts not only improved the germination of cauliflowers when applied as a seed dressing but also decreased populations of harmful microorganisms.[11] Interestingly, they also found that the extracts that gave the best microbe control – alder buckthorn (*Frangula alnus*) and maize (*Zea mays*) – were also those that gave the worst germination. This is positive news for those of us hoping to take advantage in a chip. It is possible that those species that will reduce our weed germination might also help by reducing pests and disease. We also need to be careful though not to get too hopeful, as effects seen in the lab are not always replicable in the field, where a wide range of influences are at work.

A review of the woodchip from five species of trees looked specifically at their potential to suppress weeds.[12] The study aimed to look at those chemicals that were water-soluble and would be likely to be released from the woodchip within forty-eight hours. This does mean that to get the best effect you might need to use freshly chipped wood (as we will see in a bit with the willow woodchip trial for apple scab). However, it is possible that there are other chemicals that are released more slowly that might also have some effect. The results from this experiment using several tree species including red maple, magnolia, swamp chestnut oak, red cedar and neem suggest that 'organic mulch with allelopathic potential can be effective for weed suppression in the field.' As with so much of our knowledge around these complicated interactions, far more research is needed.

Effect of Mulching on Different Weed Types

Here we look at how mulching works on four types of weed. This is not an exhaustive list of all types of weed; however, from my experience, this illustrates the potential and the drawbacks of using woodchip mulching for long-term weed control in uncultivated situations. Using it in vegetable growing and no-dig systems is covered elsewhere.

ANNUAL WEEDS

These are weeds that grow and seed in one year. They usually colonise bare ground very quickly and grow swiftly. Examples include fat hen (*Chenopodium album*), groundsel (*Senecio vulgaris*) and annual meadow grass (*Poa annua*). Mulching with a thick layer of coarse chip works well on annual weeds in the first year after application. It prevents germination of seeds already on the ground and ones that land on the mulch. From year two, it becomes less effective as the chip breaks down to a finer texture, perhaps losing some of its allelopathic qualities. Though this varies a little depending on the tree species of the mulch used, it is likely that for effecting long-term control of annual weeds a yearly top-up of mulch will be needed. Alternatively, occasional raking or cultivation of the surface will kill off any seedlings. Though this might seem to be missing the point of the weed control, it might still be much quicker and easier than if no mulch had been applied.

CREEPING SURFACE WEEDS

These types of weed are not a major issue in all systems; for instance, amongst trees or large perennial shrubs, they will not compete with your plants and may provide a good ground cover that prevents other more problematic weeds taking hold. I'm talking about things like creeping buttercup (*Ranunculus repens*) and creeping charlie (*Glechoma hederacea*). Where they cause the biggest difficulties is at the edges of systems, such as path edges or where trees and shrubs are grown with annual and herbaceous flowers and vegetables.

Mulching with woodchip can help to control them, though it is very hard to eliminate them entirely. A very thick layer of mulch will certainly slow them down and may even stop them growing through initially. However, there is always a bit of plant or seed left and this will quickly re-establish them. They mostly grow very happily in the mulch, especially as it degrades. The one advantage is that pulling them from a mulch is much easier than from most soils, so weeding becomes quicker. A concerted effort of mulching and weeding can reduce their impact, or in a garden and allotment situation might eradicate them.

NON-RHIZOMATOUS
LOWER-GROWING PERENNIAL WEEDS

I'm thinking here of weeds like dandelion (*Taraxacum officinale*) and broad-leafed plantain (*Plantago major*). These can be a significant challenge as they have deep roots and are hard to dig out. The good news is that woodchip mulch is very effective against these types of weeds. Though it may take a while to kill them off entirely, the rosette-forming nature of their growth means they struggle to push through the mulch.

RHIZOMATOUS AND SUPER-VIGOROUS PERENNIALS

Many of you will have had long-standing battles with these weeds and have the scars to prove it; docks (*Rumex obtusifolius*) and creeping thistle (*Cirsium arvense*) are those that I have most experience of, though couch grass (*Elymus repens*) is a major issue in US orchards. They have deep roots, big reserves of energy and strong shoots that can push up through almost any depth of mulch. Some, like creeping thistle, can also sneak under the mulch and take advantage of the lack of competition from other weeds that the woodchip barrier

Woodchip mulch not controlling creeping thistle at Eastbrook Farm. Though not ideal, the thistle did provide wonderful food for our resident flocks of goldfinches.

has created. This lack of competition is one of the main reasons why these weeds thrive in heavy mulches. Once they get through the barrier, there is nothing to challenge them for light or moisture; the mulch ends up helping them in just the way we have intended it to do for our crops.

Seasonal and Annual Variations

As you would expect from any management technique, there will be differences in effectiveness from season to season. In a dry year, annual weeds will be less likely to germinate on the surface of a mulch than they would be when periodic rain is keeping the surface wet. The lifespan of a mulch will be similarly affected. Cool, damp conditions favour fungal activity that will result in the mulch breaking down more quickly than hot, dry weather. Taking account of these variations or reacting to them where possible will help to make the best use of any mulching strategy.

Pest and Disease Control

Though weed control benefits of woodchip mulches are obvious, what isn't so widely understood is the potential they have for reducing pest and disease levels. It seems that, in addition to the general increased health woodchip mulches can bring to soil and plants, giving them a better chance to fight off attacks, there are some very specific benefits for certain problems that are being discovered. We will explore a few here as examples to what might be possible in diminishing pest and disease populations. This is of particular interest in tree and perennial crops, where use of rotations and break crops for this purpose are not an option.

First, let us look at how mulches can increase the biological diversity and activity in soils. It will come as no surprise that feeding the soil with organic matter will increase the biological activity, but what we are still learning about is how different treatments might affect the soil in different ways.

Glynn Percival, a tree and soil researcher in the UK who has done extensive work on mulches, states, 'Mulches supply organic matter to soils promoting the microbial population in the soil that inhibits activity of for example *Phytophthora* and *Armillaria* pathogens. Areas with poor soils low in organic matter are not able to suppress as effectively.'[13] In other words, high populations of diverse microbes will keep the few problem ones under control.

One study looking at composted poultry manure mulches in orchard systems found that: 'The compost significantly affected arthropod abundance during two years after application, with more predators and fewer herbivores in the compost treated plots. Populations of spotted tentiform leaf miner (*Phyllonorycter blancardella*) and migrating woolly apple aphid (*Eriosoma lanigerum*) nymphs were reduced in the compost plots.'[14]

Guy Ashburner was an avocado grower in Australia. He noticed that the pathogen *Phytophthora cinnamomi* (a cause of root rot in his groves) didn't seem to be present in the forests further down the hill from his infected fields. Like so many of the pioneers in this woodchip story, he took the logical next step of trying to recreate the local forest soil conditions in his plantations by adding mulches, mostly woodchip, as well as lime and manure. His approach, now known as the 'Ashburner system', revolutionised avocado growing in Australia and has been adopted by many growers around the world, though the exact mechanisms that make it successful are not fully understood. Downer et al. found that 'mulches increased microbial activity and cellulase enzyme activities relative to unmulched control trees.'[15] These enzymes are known to be able to break down cell walls of the *Phytophthora* and prevent its sporulation. They observed a greater breakdown of *Phytophthora* in mulches compared to unmulched soils and believe it may be this activity that makes the Ashburner system successful. This is backed up by the fact that, 'This protection is not afforded roots produced in deeper soil layers, because the enzymes are absorbed and deactivated on clay particles. Thus, although the mulching system controls the disease, control is limited to organic layers in soils.'[16]

There are other mechanisms, too, that may be playing a role in the ability of woodchip mulch to suppress diseases and pests.

Nutrient Deficiencies

I don't intend to go into too much detail on this, as the principle of feeding the soil to feed the plant is well known. Improving the biology of your soil will make nutrients more available to your plants. There are those who believe that the soil (and underlying bedrock) will provide all the nutrients that you will ever need, provided you are creating the right conditions for them to become available. Excepting the case of soils that have a particular deficiency (for instance, many soils in the UK are low in copper) this might well be the case, though in highly productive systems where we are taking a lot of nutrients out in the form of a crop, this can be difficult to achieve.

One area with massive potential for improving soil and plant health is our urban environment. Many of the trees growing in cities and towns are poorly cared for, often with little soil around them, or with short grass and heavy foot traffic leading to soil compaction. Adding woodchip mulch is a relatively simple treatment that can strengthen the trees' root systems, improve tree health, increase their lifespan and reduce the risk of them falling over. One study found that not only was soil bulk density reduced, but the mycorrhizal density increased when trees were mulched.[17] They also saw a reduction of chlorosis in the oak trees studied. Interestingly, when I asked one researcher if they had any pictures of deficiency with woodchip mulches that I could use in this book to illustrate the risk of nitrogen lock-up, they told me that although tests had showed some decline of nitrogen in the leaf, it had never shown as visible symptoms.

Allelopathy

We've already looked at allelopathy in some depth in chapter 1 when exploring the different characteristics of woodchip from various tree species. Though normally a consideration when thinking of potential risk from woodchip, it may be possible to use allelopathic qualities to our advantage in controlling disease. There is plenty of evidence that allelopathic chemicals from a range of trees can be beneficial to germination and growth of seedlings, as well as increasing transplant success.[18] Though conversely, others have observed negative effects on germination and growth from using these allelochemicals dissolved

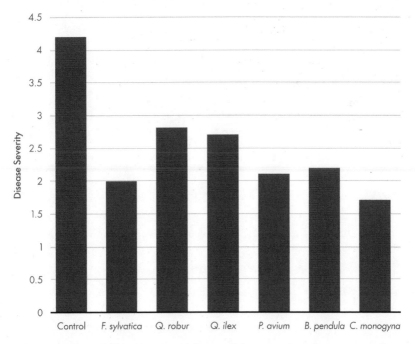

Influence of single-species woodchip mulches on disease severity of *Phytophthora* root rot. *Source:* Percival (2013).

in water. A reminder about how little we still really understand the complex interactions happening in the soil.

Glynn Percival in his work on single-species mulches found some interesting effects. For instance, he looked at how well a range of mulches suppressed *Phytophthora* on repotted horse chestnut trees.

As another example, western red cedar (*Thuja plicata*) chip releases thujaplicin as it breaks down. This chemical can affect the growth of fungi and bacteria, in situations where the balance of microorganisms has been upset. Using a woodchip that contains this compound may help to knock back the dominant harmful species, while at the same time building conditions that favour a wider range of organisms to help restore that balance. Some attempts have also been made to isolate certain allelochemicals to use as a fungicide, but it is not a well-explored area, and there is a risk of harming the very plant that you are trying to protect.

To take advantage of any allelopathic properties of woodchip it is likely that you will need to use a single-species woodchip with a known allelochemical characteristic. Though often not an option, if you have the space to grow, or the opportunity to procure, the right species of woodchip and have disease you are looking to prevent, this could be an avenue worth pursuing.

Note also that mulching can increase the risk of disease, if for instance it increases the moisture content of an already damp infected soil, providing perfect conditions for that disease to proliferate.[19] There is emerging evidence, not yet published, that some post-harvest pear diseases can increase on crops from mulched trees. In some instances, on wetter soils for example, a sprinkling of mulch may be more effective than a deep layer, if the aim is disease control or stimulation of biological activity in the soil. Obviously, this will not then give you weed control, but could be combined with other management strategies.

Examples of Specific Pests and Diseases

Having looked at the principles of how woodchip might help reduce pest and disease incidence, let's examine in a bit more detail a couple of promising examples to explore the practicalities. To be useful at scale, these techniques have to be not only effective but also workable in a farm or garden scale.

APPLE SCAB

I've talked about the Soil Association's Innovative Farmers programme in chapter 4, where we trialled woodchip propagating media. Another field lab trial from the same programme supported a group of cider apple growers to look at the potential for using willow woodchip to control apple scab. The trial was inspired by the work of Dr Glynn Percival, who noticed, as a side effect from work on single-species mulches, that apple trees mulched with willow chip seemed to have less scab on them. Scab is a significant challenge for apple growers, resulting in cosmetic damage and, in bad cases, reduced yield and serious reduction in tree health.

With few current effective chemical treatments and the likelihood of existing products being withdrawn from market, growers are keen to try other ways to reduce the disease. The trial looked both at the efficacy of the mulch and the practicality of applying woodchip. Glynn believes that the salicylic acid in the willow wood bark triggers an immune response in the apple tree, making it better at fighting

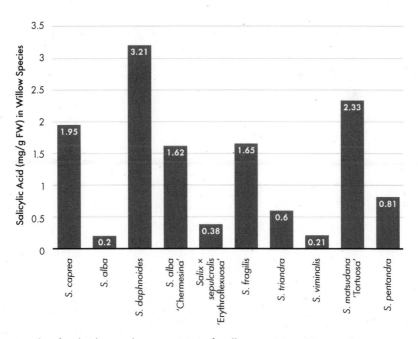

Levels of salicylic acid in a range of willow species. *Source:* Innovative Farmers 'Willow Woodchip for Top Fruit Scab Control' Field Lab report.

diseases. Similar effects have been seen in tomatoes when spraying the leaves with soluble aspirin, but using woodchip to induce the response on a large tree is another matter.[20] It is important to harvest and use the chip at the right time; levels of the acid are highest just as the sap starts to rise in early spring and quickly leach out of the chip, so it must be used quickly.

During the trial we also analysed a range of different willow species to find out what levels of salicylic acid are present and found a surprising range.

We couldn't get enough woodchip from species with the highest levels, but further trials should concentrate on these ones.

The results from the trial showed a trend across all the sites for lower scab and higher leaf nutrient content, though the results were not significant. The team observed that some sites had not mulched as heavily as recommended; this was due to concerns about mulch residues interfering with the mechanical harvesting of the apples. Thicker mulching and the varying levels of salicylic acid in the varieties leave open the possibility to further explore this as a method to reduce scab. Glynn also suggested that with the levels of salicylic acid being higher in the bark, using only that part of the plant might give stronger results. The bark is a by-product from the weaving industry, so might be available to some growers.

For commercial cider growers, applying large amounts of mulch is problematic at harvest time, but for fruit that is handpicked this should not be an issue. Growing your own willow to complement fruit systems could be a genuine opportunity to reduce sprays, or to find options to control scab in unsprayed orchards.

CODLING MOTH

There are numerous natural predators of codling moth, though it is not always possible to measure the extent of the predation since the predators eat the evidence, and many are active at night, making it harder to monitor their activity.[21] The perfect murder! Lots of birds eat them, as do beetles, spiders and mites – even rodents and bats are thought to play a role. The other major player is parasitoids, with over one hundred species identified that feed on codling moth.[22] All of this is a strong argument for enhancing biodiversity generally in orchards.

One study looked specifically at how woodchip mulches can be used to create a habitat in orchards that can be manipulated to control the moth.[23] They discovered that mulches such as woodchips provided a great habitat for overwintering pest larvae. This enabled them to treat the woodchip with a biological control of parasitoids (specifically the entomopathogenic nematodes, *Steinernema carpocapsae* and *S. feltiae*). The woodchip provided a habitat for the nematodes that was easier to control with irrigation and therefore gave a better control. Though this may seem like an awful lot of effort to go to in controlling a pest, it is better than losing crops. Remember, too, that in addition to the pest control, you are getting the water savings, weed control and soil improvement benefits. Codling moth is not easy to control by spraying, even with chemical treatments, and for organic growers is almost impossible.

Temperature

Most plants will not thrive in extreme temperatures, even if they can cope with and survive them. Similarly, soil organisms are most active in temperatures of around 25 to 30 degrees Celsius, though fungi can cope better with lower temperatures, while bacteria are less affected by high temperatures.[24] It is in our interests as growers to reduce the stress on soil biology and plants by mitigating the extremes of temperature. Using a woodchip mulch is one way to moderate both daily and seasonal temperature fluctuations.[25]

In colder climates, temperatures in unmulched soils are as much as 10 degrees Celsius higher than in similar mulched ones. A 6 to 8 centimetre (2 to 3 inch) thick organic mulch can keep soil temperatures up in the winter. This allows the roots to keep growing for longer, leading to more productive trees.[26] Conversely, soils under mulches are slower to heat up in the summer, though eventually they do warm up to the same temperature as unmulched soils at most depths. For climates with high temperature differences between night and day, this buffer will reduce heat loss at night and quick heating during the day. Some apple rootstocks like M9 have been shown to be more sensitive to hot soils, so being able to moderate the temperature on these could make a significant difference to the crop performance. If you really want your soils to warm quickly in the spring, Greenly and Rakow found in their study that a 7.5 centimetre (3 inch) depth 'appears optimum, because it maintains a significantly increased soil moisture level and does not significantly reduce the soil temperature over the season.' However, different sites, cropping situations and climates will have different requirements.

Studies such as this illustrate one key point on using mulches for moisture, temperature or pest and disease management. There may be negative effects as well as positive. What is advantageous in one situation may cause problems in another. Understanding what you want to achieve is crucial. My accidental experiment with a thick willow mulch layer worked perfectly in the drought of 2018, but in a wet year might have caused rotting in some tree species or slowed spring growth by reducing soil warming.

Thick mulch layers are so good at insulating the soil and reducing heat absorption during the day, as well as heat radiation at night, that they can increase the risk from frost. For trees this is not particularly an issue, but for young or tender plants, using a mulch can either cause frost damage or weaken the plant and make them more susceptible to diseases.

For pot-grown plants that are particularly susceptible to temperature fluctuations, using a woodchip mulch appears to make little difference in protecting from the heat.[27] The colour of the pot seems to have more of an effect. This is not surprising since the mulch would only be protecting the surface of the substrate in the pot, while the sides are still exposed to the sun. However, some nurseries that experience hard frosts in the winter use a woodchip mulch to cover entire areas of plants, to the top of the pot. This gives some protection to the roots and reduces damage and plant losses. It reduces the effect from the cold snow and freezing air, but also helps hold in any residual heat left in the soil underneath.

Heeling in Bareroot Trees

Many of my finest discoveries have been by accident. Late on one Friday afternoon I took delivery of more than three hundred fruit trees, bareroot and bundled. Light was fading and the ground was heavy and wet;

my dinner was calling. Properly heeling trees in usually involves digging a trench, laying the trees carefully in the trench and backfilling with the soil around the roots to ensure they don't dry out and die. This was just not an option on this occasion. However, I did have a huge pile of lovely damp woodchip to hand. So, as a temporary measure, I popped them into the pile, intending to do a more professional job when I returned the following week.

Bareroot tree saplings heeled in, in woodchip, prior to planting.

When I dug up the first bundle, I realised that they were perfectly happy in the wood-chip, in fact probably happier than they would have been in a cold, hastily dug, spade-smeared clay trench. So, I decided just to leave them there until they were planted. In all, they stayed in the woodchip for about six weeks and were just fine. Right toward the end of this time we had a rabbit invasion and a few trees had their bark nibbled, but this could have been easily prevented if I had appreciated that rabbits liked climbing woodchip mountains. Regular rain and cool but not cold weather also helped us. It is possible that in a heavy frost the roots might have been damaged. I now use woodchip to heel in all my bareroot plants that are destined for a quick plant. Trees that need to wait longer before planting I still do in the soil in the traditional way.

Mulches for Paths

We all suffer the muddy path problem. We, quite rightly, avoid walking on our vegetable beds or productive areas of the garden and stick to paths. In most cases these paths are grass. This is usually okay in the summer when the ground is hard and the grass growing, but once winter kicks in, the soil gets wetter and the grass is unable to heal itself from foot or tractor damage. Things get messy! We've seen already the potential of mulch to help protect the soil surface and prevent compaction. On paths this is even more helpful. A couple of inches of mulch will be plenty to protect against foot traffic and wheelbarrows in most situations. For quad bikes and tractors, a thicker layer may be needed. Spread the mulch while the ground is still hard, as this will protect the soil surface as the conditions get wetter. If you wait until the earth is muddy before mulching, some of the damage is already done and the path is likely to stay sodden for longer.

If possible, put a layer of something more solid under the mulch to provide a more stable base for the woodchip and extra protection for the soil underneath. More importantly, this helps prevent perennial weeds from coming up through the mulch. For smaller areas in home and community gardens, or allotments, you can use cardboard, thick layers of newspaper or old hessian sacks as a sub layer. I have seen plenty of examples where serious weed problems have been eliminated using this method. It can even control pernicious spreading weeds like couch grass (*Elymus repens*) and creeping buttercup (*Ranunculus repens*). Some weeds will push their way up through any depth of mulch, so if you have brambles (*Rubus*) or docks (*Rumex obtusifolius*), for instance, in your paths, try and dig out the worst of them before laying the mulch.

For larger areas, cardboard and newspaper become impractical and landscaping fabrics are often recommended. From my experience, in most farming situations you can manage without anything underneath. If you are driving over them, or walking regularly, this will reduce weed growth, and if you top up the woodchip annually any weeds that do come through can be kept in check with a strimmer or topper. The big concern with fabric membranes is that they are mostly made from plastic, and as they break down, they could be causing long-term damage to the soil. The fabric also prevents soil organisms and particularly worms from being able to move freely to bring the decomposing woodchip particles down to the lower layers of the soil. The final, and maybe most important drawback to membranes, is that they are not very effective. They temporarily stop perennial weeds from growing through them. However, inevitably seed from those weeds will land and germinate in the mulch on top of the membrane. Once those plants get going the roots quickly find a way through the membrane into the soil. Removing a dock or thistle from a plastic membrane is much harder than pulling it up from a loose woodchip mulch.

Depending on the width and length of your paths, consider digging a shallow trench before mulching. This can help with creating raised beds beside the path but will also get you down to slightly firmer subsoil (depending on your soil). You can then fill up the excavated path with woodchip. Since the level of the path will be the same as or lower than the bed, this reduces the amount of fresh woodchip spillage onto your growing beds. At the end of each season when the chip is half composted, you can either shift it onto the growing beds as a soil conditioner or add a new layer of mulch on top. This system works well in tunnels and glasshouses as well as outside, though the chip will likely last a bit longer in the drier protected climate. You can build path edges from wood, brick or other materials, but I have mostly found they are not needed from a practical standpoint, though aesthetically, of course, they may be worthwhile additions.

Cost of Spreading Woodchip Mulches

If your main concern is building long-term soil and plant health, or if you are working on a small scale or garden scale, you might be less concerned about how much you spend spreading the mulch. However, in commercial settings we must weigh up a range of approaches to weed control and soil management and get an idea of cost versus benefit.

The most comprehensive analysis of the financials of woodchip mulches compared to other systems of weed control that I have found is the superb work of David Granatstein at Washington State University Center for Sustaining Agriculture and Natural Resources Wenatchee, which we will look at on the following pages. Exact financial comparisons depend on local conditions. The most striking thing you will see, however, is that while applying woodchip mulch is expensive, and seems particularly so when compared to herbicide control for instance, there is often an indirect profit advantage within only a couple of years through improved soil and plant health. David's research showed improved fruit size and yield that provided a clear return on that investment in chip.

Costs

Material cost. Often free if getting from a waste stream, but if you are producing your own, factor in the cost of machinery hire or purchase and time taken. Even with free material there will be time and cost investment in receiving and managing the chip to the point of spreading.

Spreading cost. This is where using woodchip gets expensive. You can do it manually, in which case the machinery cost is next to nothing – a trailer and some forks – but the labour cost very quickly adds up. On the other hand, if you have the scale to justify it, you can invest in some specialist spreading kit. In the UK, to my knowledge, there is only one manufacturer of specialist compost side-discharge spreaders for

Small Whatcom spreader with woodchips mulching an apple orchard. Photo courtesy of David Granatstein.

orchards: Seymour. Kuhn and Whatcom also make a range of specialist spreaders for the US market.

For most systems, larger spreaders will be unsuitable; you won't be able to get around the trees easily and there is a risk of compaction near the tree roots. Choosing a model designed for orchards is more likely to work. For instance, the Seymour Agrofer model is only 1.6 metres (5 feet) wide.

At Eastbrook Farm, we adapted a silage spreading wagon for our wider-space alleys. We added a chute extension to get right to the base of the tree, without having to drive right next to it. The tractor was 2 metres (6 feet) from the tree row. This worked well but did rely on an even chip size. Our homemade chip went through smoothly, while the less uniform free chip from our tree surgeon, complete with occasional lumps of wood and branches, was less successful, with the unwelcome side effect of raising the blood pressure of our farm manager, who had lent us the machine. We drove slowly to ensure a discharge slow

Kuhn side-discharge spreader. Photo courtesy of David Granatstein.

Willow woodchip mulch spread using an adapted forage wagon on alley planting, Eastbrook Farm.

enough to get a dribble of chip placed where we wanted it, rather than spraying it out all over the row.

There is still, of course, a labour cost when using these types of machines, and you will need a front loader to be able to get the chip into them. However, not factoring in the setup and tweaking time, we were able to mulch a 400 metre (1,300 foot) row of trees in about an hour. We started by mulching both sides to get an even distribution of material, but with the converted silage wagon we couldn't reduce the flow of chip quite as much as we wanted. This meant that two passes resulted in too much mulch. We finally decided it was more effective to do just one pass and then quickly go up the row with a rake and manually pull the chip around the other side of the tree. As you can see from the image above, it had pretty much covered it all, so it was a very quick job.

COMPARISON AGAINST OTHER WEED CONTROL STRATEGIES

So how does this cost compare to other forms of weed control? David Granatstein has the following comparison from his work on this. You can see the cheapest controls are flaming and tillage; herbicides used in non-organic orchards would be a similar cost. Mowing is more expensive, while weed fabrics and woodchip mulch (1 metre / 3 feet wide, 10 centimetres / 4 inches thick) were more costly still, as were the organic herbicides (which would not be allowed in any case in UK organic systems).

Just looking at these costs you might decide to go for the cheapest options. However, the yield performance of the trees under these different

Table 6.1. Weed Control Costs in Organic Orchards

	$ per Acre per Year	Year
Flame weed + hand hoe	208	2014
Weed fabric (with the cost spread over a 10-year period – the estimated life of the material)	420	2014
Flaming (5 times per year)	113	2012
Tillage (5 times per year using Wonder Weeder rolling tine cultivator)	115	2012
Wood chip mulch (3-year life)	400	2012
Organic herbicide (4 times per year)	508	2012
Mowing	210	2010

Source: David Granatstein, 'Orchard Floor Management' (Power-Point presentation, Organic Orchard Floor Management Workshop, October 11, 2016), http://tfrec.cahnrs.wsu.edu/organicag/wp-content/uploads/sites/9/2017/04/Granatstein_OFMr2.pdf.

techniques tell a different story. They discounted the organic herbicide treatment, as it gave poor control of weeds and was so expensive.

The younger Gala apple trees (eight-plus years old) showed an immediate yield increase under woodchip. This difference increased in the subsequent two years, such that in year three there was a more than two-thirds improvement in yield. The more mature pear trees took time to respond to the woodchip mulch, but even they saw an increase in year two. It likely that this yield benefit is down not just to the weed control properties of the mulch but to the wide range of other potential benefits discussed during this chapter: water conservation (though plastic mulches will also provide this) and biological effects. The p (probability) value comes from an Analysis of Variance (ANOVA) and is, very crudely, a method of assessing the significance of a set of results. The closer to 1 the p value is the less significant the results, while a p value nearing zero will give us more confidence the result is showing true differences between the samples.

Soil type and crop also affect the benefit of using woodchip. For instance, an Australian blueberry study looking at replacing plastic

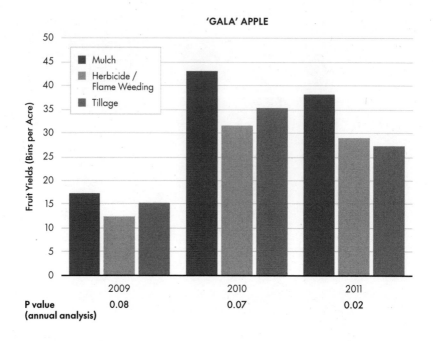

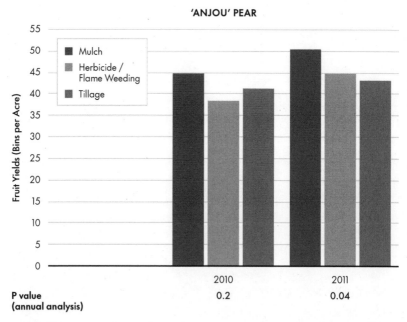

Yield effects on Gala apple and Anjou pear of a range of weed control strategies in commercial organic orchards, with large-scale field plots. *Source:* Granatstein (2014).

Table 6.2. Grower Returns with Different Weed Control Strategies

	2009	2010	2011	Three-Year Total Relative to Tillage
Apple Returns* ($ per acre) 8+ year old 'Gala' Apples / M26, Sandy Soil				
Mulch	$2,320	$8,440	$12,764	+$4,777
Herbicide / flame weeding	$1,971	$6,193	$9,638	−$946
Tillage	$2,942	$6,843	$8,963	—
Pear Returns* ($ per acre) Mature 'd'Anjou' Pears, Good Soil				
Mulch	$9,580	$12,636	$9,377	+$1,432
Herbicide / flame weeding	$10,274	$10,621	$8,141	−$1,125
Tillage	$10,676	$11,182	$8,302	—

* Gross bin returns minus weed control costs and picking costs

Source: David Granatstein, Preston Andrews and Alan Groff, 'Productivity, Economics, and Fruit and Soil Quality of Weed Management Systems in Commercial Organic Orchards in Washington State, USA', *Organic Agriculture* 4 (2014): 197–207, https://doi.org/10.1007/s13165-014-0068-0.

mulches compared their standard films against a 5 centimetre (2 inch) deep woodchip mulch.[28] After three years of the trial, they found that in the 'grey gravelly silty loam' soil the woodchip treatment increased berry yield by 81 per cent but had no significant effect on the other two soil types tested (red loam, red silty loam).

Are these increases in output, however, enough to justify the significant extra cost? Here the study compared the woodchip mulch and herbicide/flame treatments to the grower returns from the most common organic method of tillage. The results are striking. In the apple orchard, the initial small increases in yield did not outweigh the cost, but by year two they were a way ahead. In year three, this profitability was compelling. Over the three years of the trial the woodchip mulch gave nearly $5,000 higher grower returns than tillage, while the mowing and flaming was lower. In the pears, we see a similar but smaller effect.

After the second year of the trial, the weeds in the woodchip started to increase. One grower experimented using a flame weeding of the top of the mulch to extend the life of the mulch; or you could cultivate or use a herbicide to give a similar benefit. If you already had the tillage equipment you could potentially combine techniques, using two years of mulch to get the yield benefits and soil health improvement, followed by a year of tillage to incorporate the composted woodchip and reduce any weeds that were starting to come in. Tillage was the most effective measure against voles.

Other studies have looked at this issue, too, and it's clear that the convincing results from Washington may not always be replicated, however other studies have shown the same trend. For instance one study of ten different mulching treatments concluded that: 'Our economic studies indicate that, for some fruit varieties, the increased crop value in mulched trees probably justifies the greater costs. Conversely, reduced fruit quality and lower pack-out crop values for trees in some herbicide systems may nullify the cost savings of the Ground Management Systems. The anticipated long-term benefits of increased soil fertility under mulches certainly are of some value, but were uncertain and inconsistent after four years in our field tests.'[29] They also thought that the benefits for smaller producers were clearer, where woodchip was cheaper and easier to get hold of, than for large-scale orchards who might struggle to obtain the volume of chip required. Some orchardists in the US have chipped their old fruit trees instead of burning them and used the chip as a mulch for the new trees, so far without any signs of disease, which is perhaps the best way to close the cycle.

In my experience, there is a clear financial benefit to mulching trees and shrubs when planting, particularly in situations where irrigating the crop is not an option. While spraying or mechanical weed control will be cheaper, in dry conditions they give no protection against soil drying; indeed they will exacerbate the situation by removing the shading effect of a ground cover and increasing evaporation from the soil surface. In wetter years, this difference will be lessened.

There is some evidence to back up my own observations. Chalker-Scott et al. looked at the survival of a range of species in a restoration site planting.[30] The 20 centimetre (8 inch) woodchip mulch performed much better than the glyphosate comparisons; however, in addition to

Using a powerful flail mower to chip cherry prunings, which can then be brushed across to mulch trees. Photo courtesy of David Granatstein.

better survival and establishment, they even found that the snowberry (*Symphoricarpos albus*) under woodchip started producing fruit earlier than the herbicide treatment. The implication, if replicable, would be that successful mulching could result in earlier cropping from tree and bush fruit with the associated improved profitability.

It is crucial, though, to monitor any strategy. One blueberry grower in the UK who pioneered woodchip mulching and saw some initial powerful results has since abandoned the practice, saying that applying it seemed to have reduced the nitrogen availability and raised the pH of the soil.

David Granatstein has an interesting take from his extensive work in this area and shared his reflections with me.

I now lean toward a thinner mulch and complementary weed control. My gut tells me you might get about 80 per cent of the mulch benefit (other than weed control) with 20 per cent of the mulch. This would make the shortage of mulch less of an issue if mulching were being adopted widely. Also, I think it could make it more profitable. For instance, we see growers using mow and blow of the alley grass, with a side delivery mower, plus flailing their prunings and sweeping them onto the tree row instead of leaving them in the alley. This essentially creates composting in the tree row – green and brown – and creates a thin mulch layer that I showed to increase young tree growth by 20 per cent compared with bare ground.

———————

Designing any mulching approach is about using your resources in the most efficient way for maximum benefit. The thick layers of mulch we have been applying at Eastbrook Farm are great for establishing trees, but would not be financially viable (or perhaps desirable in other ways) as an ongoing strategy. However, thinner or less frequent applications, combined with other methods, could offer a good way forward.

CHAPTER SEVEN

Mushrooms in Woodchip

Woodchip without fungi is like a sea without fish: just wrong. You can't leave a pile of woodchip for long before a mushroom comes looking for a home and takes possession of it. This is one of the reasons that woodchip is so useful to us as growers and farmers. When using woodchip as a mulch, soil additive or animal bedding, we mostly are not too fussy about which fungi colonise our heap but are happy just to welcome any species that it is doing the job of breaking down the woodchip. Are we missing a potential output from our woodchip? Mushrooms are a welcome addition to the cooking pot or could even provide an additional farm income.

Several specialty mushrooms grow well outside in woodchip. With some investment and effort, you can inoculate woodchip with commercially available spawn. This can work at small scale in a home garden, either in mulch or in bags or buckets of woodchip.

There are numerous market opportunities for farm-grown mushrooms, though, as with any commercial enterprise, if you are going to do it properly you need to invest time and money to ensure a reliable commercial crop. My initial experiments have demonstrated that it is not as easy as simply throwing some spawn at a pile of chip and walking away. If you have your own retail outlets, they can be a useful addition; however, many box schemes, farm shops and restaurants are also on the lookout for alternatives to the ubiquitous button mushrooms. There's also research and trial systems into growing the tricky species like truffles and morels, which command a high price to specialist markets, though these are beyond the scope of the low-tech approach I am suggesting here.

Straw, sawdust and logs are other good substrates for mushroom farming, and while some varieties will do fine on a range of growing media, others are a bit more particular. We'll have a look at which offer the best opportunity for woodchip cultivation.

Alternatives to Growing in Woodchip

Though we are concerned in this book with what mushrooms we can grow on woodchip, it is worth just considering some of the other substrates that can be used for growing edible fungi.

The most obvious is **logs**, since if you are considering growing mushrooms as a commercial enterprise you will need to compare the cost of chipping, collecting and stacking woodchip against just using the unprocessed logs. Not surprisingly, since mushrooms have not evolved alongside petrol-driven chippers, any mushrooms that grow well on woodchip will also do just fine in a whole log. However, the fungus takes longer to colonise a log; the mycelia have an easier job pushing through the already broken-up chip. Woodchip beds are also likely to finish fruiting more quickly than logs. For some species, such as *Stropharia rugosoannulata*, the spent chip can be easily incorporated into a new woodchip production bed each year to give you a new crop, though there is a risk of reduced vitality and yield. For this reason, it is best to bring in new spawn occasionally.

Sawdust is an excellent mushroom substrate. It can be very detrimental to soil health when used raw due to its high carbon to nitrogen ratio and fine structure. However, partly decomposed by mushrooms, spent sawdust material can be safely incorporated into your woodchip pile for mulching or other soil health uses.

Some species grow best on **straw**, such as oyster and *Agaricus* species. Once pasteurized, straw is easily colonised (whereas hardwood logs or chips put up a little more resistance), furnishing you with a fast flush of fruiting bodies. *Agaricus* species, which include traditional button mushrooms and field mushrooms, are one of the few that don't do well on woodchip. Though not essential for getting a crop of mushrooms, commercially lime and manure are often added to straw to get maximum yields. One study found that rice straw with 10 per cent manure (poultry or horse) and 1 per cent lime gave the best results for oyster mushrooms. The aim is to reach a nitrogen content of 1.5 per cent.[1]

Coffee grounds are a potential substrate for some species, most notably oysters and shiitake. They have a couple of major advantages,

Table 7.1. Suitability of Substrate Type for Mushroom Species

Substrate	Suitable Mushroom Species
Woodchip (sawdust)	Stropharia, reishi, lion's mane, shiitake, maitake, oyster species, morel
Fresh logs	Reishi, lion's mane, shiitake, maitake, oyster species, turkey tail
Straw	Agaricus species, oyster species
Manure	Button mushroom
Coffee grounds	Shiitake, oyster species

being a (usually) free waste product and if used within twenty-four hours of being used for making coffee they are already sterilised. Larger production systems using coffee grounds need a bulkier ingredient added, since the grounds can become compacted from their own weight and the mycelium struggle to get enough air to thrive. There is no reason why you cannot add coffee grounds to your woodchip mix if you don't want to create a purely coffee-based system.

As you can see from Table 7.1, many species of edible mushroom can be grown on woodchip. A note of caution, though: for many of these species the woodchip must be pasteurised to ensure colonisation and a high spawning ratio. There are some that are 'entry level' species, such as oyster and Stropharia, that are suited to open beds and lower expertise. Others, such as shiitake, can work well on a mixture of ingredients, such as coffee grounds and sawdust, where the coffee gives a boost of nitrogen. And, of course, the above list of substrates is not exhaustive; wheat or rice bran, for instance, are worth adding.

There are some fungi that grow in very particular natural habitats which are difficult to emulate. Morels are an example of that, though there is growing interest and research into cultivating them. They are very hard to cultivate indoors, but recreating their favoured habitat with woodchip from suitable tree species seems to have potential. In nature, they tend to grow around deciduous trees such as elm, oak and ash, and can even grow under old apple trees. Unlike many woodchip-suitable varieties that grow directly on the wood, morels grow in the soil. However, like all cultivatable mushrooms they require the nutrients from dead wood and leaf materials to grow. They appear in spring as the dappled sun penetrates the tree's leafless canopy and warms the soil and do best when there is a cool, damp start to the year.

What Mushrooms Need

Though we have considered elsewhere how wood is broken down by mushrooms, let's take a brief look from the other end of the lens to examine what mushrooms need to thrive. This is a general guide designed for outdoor production, aiming to create conditions that in most years will be suitable for mushroom production. If you are looking for a more controlled production system with woodchip, you will need to research the more exact requirements for the relevant species.

Temperature

The mycelium of fungi has evolved to sit out cold winters, so can enter a dormant stage when temperatures drop. They start growing again as the temperature increases. For most species a minimum of 10 degrees Celsius (50 degrees Fahrenheit) will get them going, and they should keep growing provided the temperature doesn't get above 30 degrees Celsius (85 degrees Fahrenheit).

Humidity

Humidity is as important as temperature for mushrooms. Unlike plants, which can regulate moisture loss by closing the stomata in their leaves, mushrooms rely on the air's relative humidity to stay healthy. Most species love it at 70 to 90 per cent humidity. Since we can't tightly control humidity as we can in indoor systems, outdoor we need to create an environment that will keep the mushrooms as moist as possible. In practice this means minimising direct sunlight and reducing wind speed. Woodland and agroforestry systems are great places to provide these conditions, and of course the woodchip itself will hold moisture and protect the mushrooms when it is not actually fruiting, as well as gradually increasing the soil organic matter where you are growing them, which enables the soil to retain moisture. In dry climates, you may need to irrigate. Use rainwater for irrigation, as tap water contains chlorine and other chemicals that can harm mushroom production.

Food

There are three groups of fungi: **Saprophytic**, which feed on dead material; **Parasitic**, which feed on living material; and **Symbiotic**, which work in partnership with a tree species, such as truffles. For our purposes we are mostly looking at the first group, and of course specifically those that like a diet of chipped wood. There are some tree species that suit particular fungi as we will see below, but to maximise our time and effort we might supplement the woodchip with other nutrients or treatments to increase the health and yield of our crop. For instance, Jaime Carrasco et al. in their 2018 study found that, 'The addition of external nutrients increases the productivity of some low yielding mushroom varieties.'[2] They also looked at the emerging techniques of using what are termed 'mushroom growth promoting' (MGP) fungi and bacteria. These are said to have the potential to stimulate growth and fruiting, as well as reducing competition from unwanted fungi in the substrate.

While these supplementary methods may be of minor interest for hobby systems, there may be scope to increase the productivity of some of the marginal species and therefore find niche varieties that others have not yet found a way to grow profitably. There is also potential to make use of waste products from elsewhere in our system or locality to supplement our woodchip. Some of the materials identified as promising in the Carrasco study were cereal meals and brans, chicken manure, cottonseed meal, grape pomace, feather flour or defatted meals from dry nuts. Other animal manures are useful additions, too, to add low levels of nitrogen to the chip as part of the mix.

Harvesting and Storage

Once you have grown your mushrooms, you need to pick them and keep them in good condition until you are ready to eat or sell them. Being rather delicate, they need some careful handling. Here are just a few things that will help you make the most of your crop.

Harvesting

Mushrooms tend to fruit in spring and fall. They are dormant in very cold and in hot, dry conditions. Often, they will fruit after rain, particularly after long dry spells. Though very little studied, they may also have a photoperiodic response. This is where organisms can respond to day length. Fungi have both photoreceptors and circadian systems, which are the two requirements for photoperiodism.[3] Though exact mechanisms are not yet fully understood, it seems that temperature has a strong influence on fungi even where photoperiodism is happening.

Choosing what size mushroom to pick depends a little on your market and available time. You will want to maximise your harvest by letting the mushrooms get as big as possible, but equally won't want to let any go over and spoil. Hugh Blogg, who developed a coffee ground bag system in the UK, advised that: 'Rather like tree fruit, some species and systems respond to an initial thinning. We found that if too many mushrooms were pinning we would thin to perhaps 5 to 8 mushrooms for an 8 kilogram (17½ pound) bag to maximise their size.' Broad recommendations are to wait until the cap has just begun to open or flatten, but before they have started to shed their spores. Check each species for optimum size and shape. For instance, king stropharia (*Stropharia rugosoannulata*) arguably tastes best before the cap has opened, but you will get lower yield by harvesting them at this stage. If you can harvest daily, you can monitor this more closely than if you are only able to get to your fungi field less regularly.

The damp conditions required for mushrooms are also perfect breeding grounds for slugs, and, yes, you guessed it: slugs love eating mushrooms, and particularly the tiny pinning mushrooms. This can give the illusion that you are getting no fruit whereas the reality is simply that the slugs beat you to it. Growing containers will reduce that risk compared to cultivating in the field. This is another factor to consider when looking at how seriously to take your mushroom crop. Are you happy to take a sporadic low yield for a system that looks after itself? In extensive woodchip-based systems, don't harvest all your mushrooms. Let some spawn for future harvests, though you can also use one year's colonised woodchip to inoculate new areas.

Storing

Mushrooms are about 90 per cent water and not very good at holding onto it. Once picked and severed from their parent mycelium they will very quickly dry and spoil. However, rotting can also be an issue, so it's best not to pick when the mushrooms are damp from rain or recent misting if you are planning to store them. Some species can also discolour as the enzymes oxidise. While this doesn't affect their edibility, it can reduce their value. If you are growing for your own pot, just get them into it as soon as possible. For a commercial enterprise you will need either a very short supply chain or cold storage of some sort. Reduce the temperature to 1 to 4 degrees Celsius (34 to 40 degrees Fahrenheit) as soon as possible after harvesting, ideally within five hours. For some species you may be able to keep them like this for more than a week.

For longer storage, consider drying or even canning and pickling. Hugh Blogg found that the young mushrooms were particularly good for pickling with herbs, oil and vinegar, and sold well at Christmas markets. This could also be a good way of using the small mushrooms removed if thinning a crop. Older mushrooms may not dry so well, as the stems can become woody and tough; however, creating mushroom powder for stocks and soups is another option. Though all these methods add layers of production, they allow you to use or sell your harvest throughout the year.

Best Species of Tree Woodchip

Growing mushrooms on conifer woodchip is possible, but not normally recommended for most species. If pre-treated with another fungus such as *Ophiostoma piliferum,* which is used in the pulping industry to control pitch and resin acid problems, conifer woodchip can successfully grow oyster mushrooms.[4] While this shows promise as a way of dealing with coniferous woodchip waste, it is a little complicated for those of us looking for a simple system of mushroom production in our farms and gardens. For this reason, stick to predominantly hardwood chip for most mushroom species, though a small percentage of conifer in a mix should not affect yields too much.

There are exceptions; *Agaricus augustus* is a mushroom that some claim was named after the Roman emperor Augustus, for whom it was supposedly a particular favourite. More likely it gets its name from the fact that it often fruits in August, or a little earlier. Of course, that month gets its name from the same Roman, so it tracks back to him eventually. The prince mushroom, as it is also known, prefers softwoods to grow on.

What about the range of hardwood species? If you have the luxury of obtaining a single-species woodchip it is worth trying to match up those mushrooms that grow particularly well on those trees. As an extreme example, *Ganoderma tsugae,* a type of reishi mushroom, will only grow on the *Tsuga* (hemlock) tree, hence its name. Many species of fungi have preferred tree hosts and pairing them up will give you better results.

In most cases, though, we will be using what woodchip we can get hold of or harvesting mixed species from our own hedges and woodland. Though you can estimate roughly what its broadleaved versus conifer ratio is by looking at and smelling chipped wood, it will be almost impossible to know which species the chip is from unless you either cut it yourself or can ask the arborist who supplied it. One of the advantages of including mushroom production within an agro-forestry system is that we can not only create the right environment for growing fungi but also grow and harvest particular species to suit our production.

For most mushroom species, the harder woods seem to be best. Oak, ash and maple, for instance, all grow a wide range of fungi. Softer broadleaved species like poplar, willow and cherry are also useful, though they may not provide such a high yield as their wood is broken down more quickly. As a rule, the quicker the tree grows the quicker fungi can break it down. There are some hardwoods that are not so suitable; walnuts and black locust are a couple that have high levels of volatile chemicals in the wood that seem to inhibit growth – though even walnut has a mushroom, lion's mane (*Hericium erinaceus*), that will grow on it. Most advice on species suitability appears to be based on growing mushrooms on logs and may not apply directly to those grown on woodchip. As with most of my life's endeavours, I always spend time researching existing knowledge but am prepared to then ignore it all to trial something new.

Methods for Growing
Mushrooms in Woodchip

Growing mushrooms on woodchip can be done under trees in the field or garden, or you can put the chip into containers and grow either inside or outdoors. Scale of production, available facilities, and time at one's disposal are things that might affect your choice of system. By putting your chip into a container, you can have more control over the production process. You might want to sterilise the chip, keep it isolated from potentially competitive fungi species or water your mushrooms. If growing inside, you can monitor and control temperature and humidity, which will help you to get the highest yield. I have come to it from a farming perspective, looking to bring an added enterprise into agroforestry systems, so am more interested in effective, low-effort production under trees, but I will consider both systems here.

Outdoor Systems

For this system, the best variety to start with is king stropharia (*Stropharia rugosoannulata*). It is vigorous and tolerates a wider range of conditions and tree species than other mushrooms. It is also one that particularly likes growing in woodchip.

The first thing is to find the best site in your garden or farm for cultivating mushrooms. They like shade, so north-facing areas, or those directly under bigger trees, are the most obvious opportunity. If you don't have the perfect spot you can compensate by creating temporary shade and by irrigating if necessary, but all this adds time and cost to the process.

Fungi hate being disturbed, so try to find an area that won't get people or animals going through it. If you have any livestock, especially chickens, you'll need to keep them off the area. You may need to fence the area or lay some wire netting on top of the woodchip until it is established. One planting we made got dug up by marauding gangs of rabbits.

Mushrooms need some light to fruit successfully; they just don't like direct sunlight. I recently tried establishing king stropharia between young trees in an agroforestry alley system where the rows of trees are only one tree thick. This species is reputed to be more tolerant of sunlight than some others, and though it may well work successfully once the trees are bigger and casting more umbrage, we had poor establishment, and I think the ground just wasn't getting enough shade for them to get off to a good start. The direct sunlight also meant the chip became too warm and desiccated.

There are various suggested methods for preparing the bed, but most have some common advice:

- Strip any growth or humus layer down to bare soil. This may involve scraping off grass or annual weeds, or even digging up any vigorous perennial weeds like docks or thistles from the plot.
- Next, lay down some layers of cardboard on the bare soil. This is the stage to add your spawn, available either on sawdust, straw or dowel pegs.
- Then spread a layer of woodchip onto the cardboard and spawn.
- Some methods suggest further layers of cardboard and woodchip.

The best time to do this is either in the spring, for a first fruit in the following autumn, or in the autumn with the expectation of a crop in the following spring. Summer conditions are too hot and dry for the mycelia to successfully colonise the woodchip. Winter temperatures, conversely, are too low for good growth. The moist cooler weather in spring and autumn provide the right growing conditions.

A note on the woodchip: when growing outside there is an increased risk of other fungi colonising your substrate. This has two risks. Firstly, the other fungi species may be more vigorous than the one you are cultivating, meaning that you'll get poor, or even zero, establishment of your chosen species. This is a big risk for less vigorous mushroom species but can even be a problem for *Stropharia*. The other, smaller but potentially more disastrous, risk is a similar-looking species getting established in the woodchip and being harvested by mistake with your cultivated variety. There are a couple of steps you can take to reduce these risks:

1. Use fresh woodchip which doesn't yet have any fungi growing on it. Chip that is less than two weeks old is unlikely to have any active fungi growth on it. If you know you are not going to be able to use the chip within that time, covering it with a tarpaulin will reduce, but not eliminate, the risk of airborne spores landing on the chip.
2. Sterilise or pasteurise the chip. Most controlled production methods recommend sterilising the growing medium before introducing spawn. Done properly, this kills off existing fungi and harmful bacteria within the substrate, providing a blank canvas onto which your chosen species can create their productive masterpiece. This is relatively straightforward to achieve on small batch production, for

instance in the bag and bucket examples below. It is much harder to do when growing large areas.

The usual methods are: **hot bath** – soaking in water at temperature of 70 to 80 degrees Celsius (160 to 180 degrees Fahrenheit) for an hour or more; or **hydrogen peroxide** – soaking the chip in a solution of 1 litre (1¾ pints) hydrogen peroxide per 4.5 litres (1 gallon) water for an hour. With the right equipment you can also oven- or steam-treat the substrate, but this is more difficult for most of us. With the right weather conditions, you could also try solar sterilisation, covering the chip with a clear plastic sheet and hoping that the temperature gets high enough.

My instinct is that for larger areas, the cost, energy and hassle involved with trying to clean the substrate are not justifiable, and that we need to make it work more naturally. So, if possible, use the freshest chip you have.

Once you have successfully established an area of mushrooms on woodchip, you can use that to inoculate other areas. Any paths or mulching you are doing can have some of the colonised woodchip added to it, with the hope of having mushrooms popping up all over the place.

Note: Growing mushrooms outdoors means that there is always the possibility of wild colonisation by potentially toxic species. Always ensure that any mushrooms eaten or sold have been positively identified as safe. If you are not sure, consult an expert.

Mushrooms in Buckets and Bags

One commonly used method for growing oyster mushrooms in woodchip is to use a 5-gallon bucket. Bags can work as a single-use alternative; however, 5 gallons is big enough to produce some decent mushrooms, but small enough that you can move it around, even when the woodchip is wet. Using a sturdy bucket means that it is reusable; you should be able to get many harvests from your bucket. If you are keen not to use plastic, you can also use old jute shopping bags or hessian potato sacks, though these will likely need more watering to prevent them drying out. They are also not reusable as the fungi will eat them as well as the woodchip.

The first step is to drill some holes in your bucket. This is to give somewhere for the fungi to fruit from. The holes don't need to be big; 6 millimetres (¼ inch) is fine, though slightly bigger is also not a problem. If they are too big the woodchip will fall out.

Next soak the woodchip. You can either do that in your pre-drilled bucket, by sticking the whole bucket into a larger container full of hot water. Or you can put the woodchip into any other container to soak and transfer. Then add alternate layers of spawn and woodchip into the bucket. Leave the bucket for two to three weeks so that the fungi can colonise the woodchip. It is vital to keep the woodchip moist during this period. If your bucket has a lid you can put that on, but it might also be worth covering the whole thing with a plastic sheet or bag to stop it drying out.

If they have colonised well, you should start seeing some 'pins' (or baby mushrooms) starting to form through the holes in the side of the bucket. Now move them to a shady, sheltered fruiting spot. This can be outside or in a shed or garage. Again, the crucial thing is not to let them dry out. You may need to water them to keep them producing. Water little and often to keep the chip moist but not wet.

In the right conditions you can often get a second or even third flush from the chip, however you can also use the infected woodchip to colonise new buckets, or add as a mulch in the garden where you might get more mushrooms.

Closing the Loop

Though mushrooms feed on the substrate as they grow, the spent material at the end of the harvest is still of value. By growing mushrooms on it we have effectively begun the composting process of the woodchip. In an extensive outdoor system, the partially digested woodchip will already be in place as mulch or woodpile and can be left in place. Material from an indoor system can augment your compost heap and newer woodchip piles. Alternatively sprinkle around existing plants or onto no-dig beds – especially good if you have used manure or coffee grounds in the mix.

Building Carbon with Woodchip

C arbon is one of the most important building blocks of life. Second only to oxygen in our bodies, it makes up around 20 per cent of a plant or animal's mass. Carbon on earth is in constant flux, moving from the air to plants, from plants to animals and back to plants, from plants and animals back into the air and so on. It can be 'locked up' – that is, not as a gas in the atmosphere – for long periods in various guises. Carbon is temporarily held onto in living entities, but there are more stable forms that also keep it out of the air. Crude oil, for instance, was mostly formed by the long-term compaction of algae, plants and animals and can stay deep below the earth's surface for millions of years. Wood will store carbon until it rots or is burnt, at which point the carbon flows back into the atmosphere. By burning wood in a controlled manner to make charcoal we can significantly increase the time the residual carbon will take to degrade. The story of biochar and the 'terra preta', as we shall see later, shows that charcoal can lock up carbon for potentially thousands of years.

In the context of the climate emergency, we must find ways to capture carbon and keep it out of the atmosphere. Building materials, furniture and plastic replacements all have massive potential, particularly where current alternatives are net carbon releasing products. However, within the scope of this book we are interested in soil and how we can make use of woody material to boost soil carbon stocks. There is an extensive body of evidence showing that soils have huge potential to store carbon, even if the exact scale of that potential is debated. It is also clear that poor soil management has the opposite effect, leading to farming becoming a carbon emitter. It has been estimated that we have lost 50 to 70 per cent of our soils' original carbon stocks; this equates to about 8 per cent of global carbon stocks.[1] On its own, replacing that carbon won't save the world, but it would make a significant contribution. Building soil carbon is not a quick or easy process, however, particularly in sandy soils and in areas with low rainfall.

As organic material breaks down it eventually becomes humus. This is often regarded as the final breakdown state of organic matter. Humus is extremely resistant to further breakdown and hence tends to last in the soil for many years. Only a small proportion of soil organic matter is humus, and so most of it can be lost quite quickly with the wrong management.

Soil is varied and complicated, which is one of the reasons for us to be cautious about claiming certainty in the carbon sequestration debate. However, let us see if we can draw some guiding principles from the evidence on whether woodchip is worth including in management approaches to build soil carbon. As with any single technique we must consider woodchip as part of a whole farm or garden system. Woodchip will not be a silver bullet, things rarely are, but I hope to show that however you work your soil and grow plants, woodchip can not only benefit you directly with healthier plants and soil but also contribute to our search for a cure for climate change.

How Nature Builds Soil Carbon

Nature, left to its own devices, creates a balance between carbon in the atmosphere and that stored in the world's soils, plants, animals and oceans. Human activity has dramatically disturbed that balance, but nature works hard to restore it. Here's how.

Leaves, Roots and Branches

Plants make organic matter by absorbing carbon dioxide from the air and manufacturing carbohydrates (through photosynthesis) and by taking mineral nutrients and water up from the soil into their roots. They use these products of photosynthesis and absorption to grow leaves, roots and stems. Some of this plant material gets browsed by creatures of all sizes, from leaf-nibbling insects to the earth's great herbivores. In annual and perennial plants. what is not eaten when winter comes dies down to become food for detritivores like worms, slugs and woodlice.

Woody perennials and trees build up a longer-lasting carbon structure in their trunks and branches, though eventually of course they, too, die and are eaten by fungi and other microbes. And just as plants lose parts of their aboveground mass, so they will suffer losses below ground, for instance when bits of root die, or as food for munching soil creatures. In addition, when a tree loses limbs, a similar proportion of the fine roots normally die to keep the tree in balance. Those sacrificed roots also then become part of the soil organic matter.

Not all this carbon gets or stays below ground, though; some is excreted above ground by soil creatures or released as the organisms breathe or rot.[2] Overall, though, there is estimated to be a net gain of about 3 gigatonnes of carbon per year (GT C/yr) stored by plants globally.[3] This compares with approximately 2 GT C/yr stored by the world's oceans. Provided we don't then release that carbon through soil cultivation or other destructive processes, we would continue to see an increase in soil carbon until that system reaches saturation. Starting with a typical cultivated soil this is estimated to be between 20 and 40 years.[4] Forest soils will stop sequestering once the soil becomes so biologically active that it emits the same carbon as it absorbs. We mustn't forget, though, that the trees in that forest are acting as a carbon store. A well-managed forest or woodland, where some wood is extracted for use, could potentially continue to sequester carbon, particularly where that wood is not burnt but is used in a long-term product such as buildings or furniture.

Root Exudates

I love looking at trees and imagining the mass of roots below the surface mirroring in volume what we can see as branches and trunks above ground. This is particularly striking with urban trees with little soil. I find it hard to imagine how they survive and where they find their food and water. I then imagine that network of roots pumping sugars and other organic compounds out into the soil. Plants invest nearly half of the food they produce into supporting other soil life that feed on these root 'exudates'. In return, the organisms that gather around their roots to feed release nutrients that the plant cannot produce itself. These are absorbed in solution into the soil roots. This symbiotic relationship is not only the basis for a healthy and thriving soil microbiome, but also helps to explain the potential for soils to sequester carbon beyond the materials deposited from leaves and dead branches. Around the plants' roots a thriving soil food web develops, processing and cycling carbon. The more life there is down there, the more carbon is held in the soil rather than floating around as a gas, until the soil reaches that balanced saturation point.

Fungi

Though all soil organisms contribute to the mass of carbon held and processed in the soil, soil fungi play a special role. They have been shown to be major drivers of carbon sequestration in forestry soil, and what is more, when atmospheric carbon levels increase, plants can use that extra carbon to produce more food to trade with fungi or other organisms.[5] A plant's growth is rarely limited by its ability to photosynthesise food; more often water or temperature extremes inhibit growth.[6] A 2004 study showed that elevated carbon dioxide levels resulted in a 47 per cent increase in mycorrhizal abundance.[7] Thus, if we can create a habitat in our soil that in other aspects encourages fungal growth, plants have the potential to pull more carbon out of the atmosphere and make it available for mycorrhizal fungi, as well as having plenty of food for themselves. Looking at that holistic system again, we can now see that if we plant trees to produce woodchip, we are creating a permanent planting that promotes fungal growth in the undisturbed soil beneath the trees, and the product from those trees are helping to build fungal systems elsewhere when we spread them as woodchip.

We have looked at the benefits of shifting soils toward a higher ratio of fungi to bacteria in the soil. This not only helps soil health but can play a part in sequestering carbon, too. Though the exact mechanisms are not fully understood, one theory is that the fungal-dominated soils have stronger crumbs (or soil aggregates) than those with a higher bacterial ratio.[8] This crumb protects the organic matter, making it less susceptible to breakdown and to carbon release back into the atmosphere. Fungi may also be more efficient at processing organic matter from the surface. However they manage it, there seems to be clear evidence that shifting our soils toward a higher fungal content will increase their potential to sequester and store carbon.[9]

Wood

Woody material normally finds its way into the soil when branches drop, or a tree dies and falls over. A whole branch can take some time to break down, which from a climate change perspective is good, but is of little practical benefit within most active farming systems. By chipping the wood and adding it to soil we speed up the carbon release compared to leaving a log or branch on the ground, but still hold some of that carbon in the soil for several years. If growing the chip is part of a tree management system such as coppicing, the act of cutting the wood keeps the trees in a high-growth state, therefore pulling more carbon out of the air than they would if they became mature trees. Though the exact potential for increasing sequestration rates by coppicing depends a lot on the species and climate, studies looking at coppicing found that a cutting rotation took more carbon out of the atmosphere than the same trees grown as a forest.[10] If the wood harvested is then used to hold carbon for either a long or short time or replaces the use of fossil fuels for heating, for instance, this can have an additional benefit.

We've looked at how carbon can be added to the soil in nature, so the next stage is to see how we can help nature, either by reducing soil carbon losses or by finding ways to add carbon through our choice of farming methods. There are a range of what are now being grouped as 'regenerative' techniques that can sit alongside the use of woodchip in helping to build soil carbon.

Techniques to Capture and Reduce Losses of Carbon

Cover crops play a significant part and have the advantage of not relying on external fertility, whether that is from another part of your own holding or from another farm or country entirely.[11]

Reducing tillage prevents the oxidation of carbon that happens during cultivations, as well as the disturbance to soil life and fungi.

Building longer rotations, with the implied use of long-term leys and in many cases using livestock, is a cornerstone of organic farming and has been shown to increase soil organic matter. However, this is usually not a quick process and indeed may take many years to show a noticeable increase if this is only technique employed.[12]

Remove or reduce nitrogen fertilisers. Despite being widely used to increase crop yields, nitrogen fertilisers can cause environmental damage in several ways. It takes a lot of energy to manufacture them, which contributes to global warming, though there are potentially more sustainable methods of fertiliser production being developed.[13] Since it is very difficult to apply exactly the right amount, there is usually some leaching into the watercourse, or oxidisation into nitrous oxide, a gas which is around three hundred times more damaging to the environment than carbon dioxide. Finally, observations from as long ago as 1927 are supported by more recent evidence; that, in the context of soil carbon, the ongoing use of mineral nitrogen fertilisers decreases soil nitrogen and therefore reduces soil organic matter.[14] Too much nitrogen speeds up the bacterial activity and therefore the release of carbon dioxide. Without the addition of organic matter to feed this activity, soil organic matter suffers. To maintain that high output, reduce nitrogen loss and build soil carbon, it may be possible to combine woodchip and nitrogen fertilisers to create a slow-release product, though it may not eliminate all the issues associated with using mineral nitrogen.[15]

Remove or reduce fungicides. Though particular products have been developed to target a specific disease, many fungicides have a harmful impact on other fungi that they meet, and that includes soil fungi.[16] There is variation between individual fungicides, dose rates and affected fungi species, as well as the recovery rate once treatment stops. Since we also know that fungi play an important role in building soil carbon, it is not a great leap to suppose that a decrease in soil fungi from the use of fungicides can negatively influence soil organic matter levels and the ability of a soil to sequester carbon, though this is not an area that appears to have been much studied.

Herbicides. Although not designed to target fungi, herbicides can have an impact on soil organisms, including fungi.[17] Even glyphosate, long touted as the safest herbicide, has been shown to impact fungal populations, though interestingly, it can boost some species and harm others.[18]

Biochar

I first became interested in the potential for biochar when I trialled a biochar compost for germinating squash and globe artichoke. I noticed that the seeds sown in compost with biochar added germinated a day or two earlier and grew more quickly than those sown in the same compost without biochar. This effect has been shown in other more scientific studies.[19] The theory is that plants have evolved to germinate and grow speedily after forest fires, and therefore the first plants to take advantage of the clearing left after the flames are the ones that thrive.

Leaving aside the carbon sequestration for a moment, what are the benefits to soil, plant and animal health from using biochar? The physical honeycomb-like structure of biochar means that it acts like a sponge and can hold onto a lot of water, and the nutrients that are dissolved in that water.[20] It can also provide a home to microorganisms, protecting them from extreme moisture or temperatures. Many biochars are alkaline. This can be great for acidic soils, but not so helpful if you already have a soil with a high pH.

Most work on the carbon sequestration potential of woodchip has been done using biochar. Biochar is effectively a special type of charcoal, in other words organic material that has been pyrolysed, or burnt, in the absence of air at a specific high temperature. The residual char is a very stable form of carbon that is currently being hyped as one of the saviours of climate change. I have been involved in a number of trials using biochar and believe it does have significant potential in certain circumstances. However, like any new god, we need to treat it with caution, recognising that we are very far from understanding all the impacts that using it at scale may have on our soil. For a deeper delve into biochar, I recommend reading *Burn* by Albert Bates and Kathleen Draper, but it is worth covering here, since one way to increase the potential benefit and value of woodchip for our soil, planet and potentially as an income stream for the farm, could be to turn it into char.

From my experience, and reading of some of the literature (I cannot claim to have read all of the extensive science), there are significant opportunities to use biochar for propagation and in poor soil conditions, especially when that poor soil is also subject to extremes of temperature or moisture, such as glasshouses or arid climates. There appears to be less benefit in richer soils, or those such as clay soils that have more ability to hold onto nutrients and water. There is even the risk that you end up boosting soil biology so much that there is a net loss in carbon from the soil.[21] The properties of biochar are also dependent to some extent on the feedstock that is used to produce it and the method of production.[22] We are still very much at the beginning of our understanding of these differences and their potential on performance in agricultural systems.

It is important to note that adding 'raw' or fresh char to soil may cause initial lock-up of nutrients. There are numerous studies that show potential yield reduction when you get it wrong. For this reason, biochar is normally mixed with something to 'pre-charge' it, such as compost or manure, or a liquid feed like slurry, digestate or plant extracts. One common method on a small scale is to add it to your compost heap for a year before putting into the soil.

Using charcoal as an additive to animal health is a centuries-old practice and one that is backed up by modern science. A 2019 review found that: 'The literature analysis shows that in most studies and for all investigated farm animal species, positive effects on different parameters such as toxin adsorption, digestion, blood values, feed efficiency, meat quality and/or greenhouse gas emissions could be found when biochar was added to feed.'[23] In addition to using biochar to feed animals, it can make a very useful addition to the bedding material, helping to absorb urea and reduce methane emissions.

The challenge when assessing biochar as a carbon sequesterer is that a proportion of the carbon in the feedstock is lost during the pyrolysis, often around one-third, though this varies significantly depending on method and material. Planting crops simply to harvest, turn into char and add to the soil is unlikely to help with our climate change challenge. However, the key is to look at what Albert Bates calls 'carbon cascades'. The principle is that you get more than one bite of the cherry. So, as an example, you place your biochar production next to a glasshouse and use the heat produced to increase plant production. You then feed the char to your cows for the health benefits (as well as the potential reduction in methane emissions). The char is passed out in the manure pre-charged; you mix it with other ingredients to make your plant propagation compost. Finally, after all these uses, it gets added to the soil, where it helps to build soil life in your greenhouse. In this scenario, most of the carbon emitted through burning is not wasted and there are multiple benefits from the final material.

The choice of feedstock is also a factor when it comes to assessing carbon benefit. Biochar can be made from almost anything organic: woodchips, of course, but also straw, coconut husks and sewage sludge, to name a few. Many of these materials are classed as wastes, though in reality many should be viewed as the replacement fertility for the soil in which the original crops or animals were raised. Herein lies the challenge, and one that we don't have the answer for yet. Is it better to add the material straight to the soil or to turn it into biochar first? If a suitably long cascade is available, it is possible that pyrolysing wins out, even though for most people that may not be either practically or financially possible. Even where a sustainable source of truly waste material is available, such as clearing a weed species like *Rhododendron ponticum*, the cost of charring and transporting prevents many farmers from making use of it. Growing and chipping your own wood onsite to build your soil carbon may be more accessible, even though that carbon may persist in the soil for a shorter period.

Burying Woodchip

Carbon capture and storage is one technology touted to help halt climate change. While scientists and engineers work on manmade solutions, trees are already doing exactly that. There is a theory that burying wood in lower anaerobic layers of the soil where it will not break down quickly is a great low-tech way to reduce the amount of carbon in the atmosphere. Evidence on the breakdown of woody material in landfill sites shows that less than 5 per cent of the carbon is lost.[24] However, this is hotly contested. The anaerobic decomposition of organic material which results from burial is a great way to generate methane, one of the major villains in the climate change story. There are imminent bans in the UK on burying organic wastes in landfill for exactly this reason. One proposal suggests harvesting large trunks and branches from the forest floor for mass-scale burial of wood.[25]

Methane issues aside, in this hypothesis the author appears in my view to have underestimated the impact that removing this material would have on the forest soil health and wider biodiversity. They do suggest selective thinning and collection in some situations, but even more of an issue is the sheer amount of land required to bury so much wood if the technique were used on a global scale. Looked at on a farm or garden level, though, there could well be situations, such as the wood-chip denitrification bioreactors we looked at in chapter 3, that provide an additional benefit of holding carbon under the soil. The *hügelkultur* beds perform the same double function, slowing down the breakdown of the logs, while using them to increase production, and maximise available water. It is important to note also that these systems are relatively shallow, and there will be some oxygen reaching the woodchip, unlike the deep burial proposal above.

Using Woodchip to Slow the Carbon Cycle

Though it has not to my knowledge been fully scientifically assessed, adding woodchip to soil will store carbon temporarily. Softer woods like willow quickly disappear from the soil surface when added as a mulch, though we can assume that some of that carbon will persist in the soil. Harder woods last for longer; for instance, a hornbeam mulch is still effective after two years and some of the larger pieces of wood can be seen even after three years. The biggest carbon gains will come from a holistic approach where the benefits cascade. Adding woodchip to soil might not be the best option when viewed purely through a soil carbon increase lens. If, however, we can grow the trees that supply the material in a way that provides wider benefits, such as shelter for crops and animals, or change the way we manage hedges from a yearly flailing to a ten-year coppice, the carbon capture that adding woodchip to soil either fresh, composted or as biochar starts to look like a useful contribution. If, in addition, we apply that woodchip as a mulch to help plant growth, or it helps us to reduce irrigation and fertiliser inputs, we can see that it could play a significant role in helping to mitigate climate change.

How much carbon is lost from woodchip before it even gets into the soil depends to some extent on how it is treated and managed. One of the benefits of using ramial chipped is that you apply it directly to the soil, where it is very quickly incorporated by soil microbes into the below-ground organic matter. Woodchip left stacked or composted will lose dry matter over time. 'Dry matter' is a term used to describe what is left from plants after all the water is removed. Roughly half of a plant's dry matter is carbon. This carbon loss is important financially to biomass producers as it represents a reduction in income and can be as much as 27 per cent in systems that are aiming to stop the decomposition of the woodchip.[26] When we're using woodchip as a soil amendment, though, we usually are keen to encourage the wood to rot down to make it more useful and reduce the risk of nitrogen lock-up. In this scenario it is possible to lose as much as two-thirds of the carbon in the woodchip through the composting process, which interestingly is similar to a typical loss in a low-tech biochar production system. Capturing the carbon dioxide as it is released or making use of the heat generated by the decomposition, would help to reduce the net loss of the system.

Mixing woodchip with other organic materials such as manure that have a higher risk of ammonia or carbon dioxide discharge, could help to mitigate some of that loss from the woodchip. For example, covering slurry or manure piles with a thick layer of woodchip has been shown to reduce emissions and would reduce the time needed before the chip could be safely applied to the soil, which in turn could lessen the amount of carbon lost compared to longer-term storage and turning.[27]

Summary

Using woodchip can have a positive impact on soil carbon levels. Firstly, there is the carbon content of that actual chip, which can be used fresh as ramial chipped wood, composted or mixed with manure, or turned into biochar. Then, depending on the growing system of the trees from which the wood comes, there is potential to increase soil organic matter through root growth and boosted fungal populations. Finally, the increase in soil health, as well as the benefits brought if woodchip is used as a mulch, can increase the growth of other crops and trees and help reduce inputs of fertiliser and water. This resulting resource efficiency and increased yields contribute to reducing the carbon footprint of a farm or garden. While there is still more work to be done to get the evidence on the true potential of specific techniques, incorporating woodchip into existing growing systems is likely to result in a reduction of carbon in the air and a corresponding increase in the soil.

ACKNOWLEDGEMENTS

This book, though drawing on my own experience, is also a testament to those friends that have inspired me to pursue my woodchip obsession. I hope that I have done them justice; my purpose in this book is to celebrate their work and vision.

Sally Westaway and Dr Jo Smith worked together on several agroforestry and woodchip projects while they were researchers at the Organic Research Centre. It was Sally who introduced me to ramial chipped wood and encouraged me in writing about it. Their work on hedges features in the book and was ahead of its time in laying out the commercial and climate benefits of this currently underused resource.

Dr Audrey Litterick, as well as being a fine musician with whom I have enjoyed many post-project meeting sessions, has a keen mind and wealth of experience that has more than once saved me from exposing my ignorance. Her work on composts and organic materials is superb.

In the field meanwhile, Tolly (Iain Tolhurst) is without question one of the most extraordinary growers I have met. At the time of life when most of us would be putting our feet up and taking life a little more slowly, he is still trying new ideas, tweaking his growing system and challenging received wisdom. It was seeing the biological activity of his soil with woodchip amendments that really pushed me from interest to evangelism on woodchip. Fred Bonestroo, too, is one of my woodchip compadres; though less scientific, he has an instinctive feel for soil and plants that I am somewhat in awe of.

As a novice to mushroom growing, I am very grateful to Dominic Thomas, founder of Fungusloci micro-farm, and Hugh Blogg, associate grower at Fungusloci and a Soil Association colleague, for their help in guiding me through the opportunities in this area.

There are a few other people without whom this book might not have happened. David Granatstein is well referenced in my text and, although we have not met, his work was a major driver in me writing this. He has also been very generous with his time and advice on helping me with the manuscript. Dr Glynn Percival's work on tree diseases is superb, and our collaboration on the willow woodchip for apple scab trial opened my eyes to the wider potential woodchip has to tackle specific challenges. I have benefitted too from reading the work of Professor Linda Chalker-Scott, which helped me to clarify and distil my thoughts.

I would also like to thank Helen Browning not just for her long-standing support in my work at the Soil Association and beyond, but for allowing me to experiment with woodchip at some scale on our agroforestry project at Eastbrook Farm. I am so grateful to Paul Clark who, alongside his career as a top-level squash professional, has moved more woodchip than possibly any other human alive and is still smiling. Without Paul, our success at Eastbrook would not have happened. Also Clive Thomas, whose expertise in forestry and patience in teaching me have been invaluable as I learn more about trees.

A massive thank you to Dr Audrey Litterick, Dr Andrew Walker, Sally Westaway and Jonathan Spencer for their gift of time and expertise in checking and editing the manuscript. Any residual gaps or errors are, of course, my own responsibility.

Finally, thanks to Ruth, Ivan and Jonah for their constant support and patience. Sorry kids, time to get off the computer now – I've finished writing.

NOTES

Chapter 1: What Is Woodchip and Where Does It Come From?

1. 'Composted Wood Chip Fines', Gristwood and Toms, accessed 24 March 2021, https://www.gristwoodandtoms.co.uk.

2. Ben Raskin and Sally Westaway, 'Field Lab: Amendments for Soil Health Final Report', Innovative Farmers (June 2020), https://www.innovativefarmers .org/field-lab?id=c6bb2819-56d3-e611-80ce-005056ad0bd4.

3. Jennifer H. Levy-Varon et al., 'Tropical Carbon Sink Accelerated by Symbiotic Dinitrogen Fixation', *Nature Communications* 10 (2019): 5637, https://www .nature.com/articles/s41467-019-13656-7.

4. J. Zavitkovski, M. Newton, 'Effect of Organic Matter and Combined Nitrogen on Nodulation and Nitrogen Fixation in Red Alder', in *Biology of Alder*, J.M. Trappe et al., eds. (Portland, Oregon: Pacific North West Forest and Range Experimental Station, USDA, 1968), 157–72, https://agris.fao.org/agris-search /search.do?recordID=US201301221698.

5. Suzanne Morse et al., 'Potential of Coppiced Alder as an On-farm Source of Fertility for Vegetable Production: Final Report for ONE13-187', Sustainable Agriculture Research and Education, accessed 24 March 2021, https:// projects.sare.org/project-reports/one13-187/.

6. R.J. Willis, '*Juglans* spp., Juglone and Allelopathy', *Allelopathy Journal* 7, no. 1 (2000): 1–55, https://www.allelopathyjournal.com/Journal_Articles/AJ%20 7%20(1)%20January,%202000%20(1-55).pdf.

7. D.A. Summers and J. Lussenhop, 'The Response of Soil Arthropods to Canopies of Black Walnut', *Pedobiologia* 16, no. 5 (1976): 389–95.

8. David N.-S. Hon and Nobuo Shiraishi, eds., *Wood and Cellulosic Chemistry*, 2nd rev. ed. (Abingdon: CRC Press, 2000), 414; Nick Umney, 'Corrosion of Metals Associated with Wood', *Conservation Journal* 4 (July 1992), http://www .vam.ac.uk/content/journals/conservation-journal/issue-04/corrosion-of -metals-associated-with-wood/.

9. W. Garrett Owen et al., 'Liming Requirements and pH Modification for Pine Wood Chips as an Alternative to Perlite', GreenhouseGrower.com (23 July 2014), https://www.researchgate.net/publication/273946343_Liming _Requirements_and_pH_Modification_for_Pine_Wood_Chips_as_an _Alternative_to_Perlite.

10. M.B. Tahboub et al., 'Chemical and Physical Properties of Soil Amended with Pecan Wood Chips', *HortScience* 43, no. 3 (June 2008): 891–96, http://doi.org/10.21273/HORTSCI.43.3.891.

Chapter 2: Sources of Woodchip

1. Michael Houlden, 'Hedges: A Renewable Source of Energy', Farming Connect Management Exchange (November 2016), https://businesswales.gov.wales/farmingconnect/sites/farmingconnect/files/michael_houlden_normandy_november_2016_eng_final_2.pdf.

2. R.F. Long and J.H. Anderson, 'Establishing Hedgerows on Farms in California', *University of California Agriculture and Natural Resources* 8390 (April 2010), http://dx.doi.org/10.3733/ucanr.8390.

3. S. Westaway and J. Smith, 'Productive Hedges: Guidance on Bringing Britain's Hedges Back into the Farm Business', Organic Research Centre (16 April 2019), http://doi.org/10.5281/zenodo.2641808.

4. Westaway and Smith, 'Productive Hedges'.

5. M. Chambers et al., 'Hedgerow Harvesting Machinery Trials Report', Organic Research Centre (2015), https://tinyurl.com/twecom.

6. Sally Westaway, 'Ramial Woodchip in Agricultural Production: WOOFS Technical Guide 1', Organic Research Centre (2020), https://www.organicresearchcentre.com/wp-content/uploads/2020/12/WOOFS_TG1_Final.pdf.

7. B. Raskin and S. Osborn, eds., *The Agroforestry Handbook: Agroforestry for the UK* (Bristol: Soil Association Limited, 2019): 9, https://www.soilassociation.org/farmers-growers/technicalinformation/agroforestry-handbook/.

8. C. Martius et al., 'Microclimate in Agroforestry Systems in Central Amazonia: Does Canopy Closure Matter to Soil Organisms?', *Agroforestry Systems* 60, no. 3 (January 2004): 291–304, https://doi.org/10.1023/B:AGFO.0000024419.20709.6c.

9. J. Wickham et al., 'A Review of Past and Current Research on Short Rotation Coppice in Ireland and Abroad', National Council for Forest Research and Development (2010), https://www.forestresearch.gov.uk/documents/2077/A_review_of_past_and_current_research_on_SRC_in_Ireland_and_abroad_2010.pdf.

10. L.J. Williams et al., 'Spatial Complementarity in Tree Crowns Explains Overyielding in Species Mixtures', *Nature Ecology & Evolution* 1 (27 February 2017): 0063, https://www.cbs.umn.edu/sites/cbs.umn.edu/files/public/downloads/2017.Williams.Nat_.ecol_.evol_.pdf.

11. Coed Cymru, 'Shelterbelts: A Guide to Increasing Farm Productivity', https://coed.cymru/images/user/IAR%20Shelterbelts%202016%20v3.pdf.

12. Environment Agency, 'T6 Waste Exemption', https://www.gov.uk/guidance/waste-exemption-t6-treating-waste-wood-and-waste-plant-matter-by-chipping-shredding-cutting-or-pulverising.

13. Hybu Cig Cymru, 'The Woodchip for Livestock Bedding Project', https://
meatpromotion.wales/images//resources/The_Woodchip_for_Livestock
_Bedding_Project_(Final_report).pdf.

14. Robert Myers Paul, 'Use of Woodchip for Agricultural Livestock Bedding',
(PhD diss., Bangor University, 2013): 113, 259, https://core.ac.uk/download
/pdf/228910725.pdf.

15. J.R.E. Johanssen et al., 'Bedding Hygiene, Cleanliness and Lying Behaviour
for Heifers Housed on Wood Chip or Straw Deep Bedding', *Acta Agriculturae
Scandinavica, Section A: Animal Science* 68, no. 2 (2018): 103–11, http://doi.org
/10.1080/09064702.2019.1601763.

16. X. Hao et al., 'Carbon, Nitrogen Balances and Greenhouse Gas Emission During
Cattle Feedlot Manure Composting', *Journal of Environmental Quality* 33, no. 1
(January 2004): 37–44, http://doi.org/10.2134/jeq2004.0037; J.J. Miller et al.,
'Available Nitrogen and Phosphorus in Soil Amended with Fresh or Composted
Cattle Manure Containing Straw or Wood-chip Bedding', *Canadian Journal of
Soil Science* 90, no. 2 (May 2010): 341–54, http://dx.doi.org/10.4141/CJSS09053.

Chapter 3: Managing Woodchip

1. J. Rex et al., 'Investigating Potential Toxicity of Leachate from Wood Chip Piles
Generated by Roadside Biomass Operations', *Forests* 7, no. 2 (February 2016):
40, http://doi.org/10.3390/f7020040.

2. P. Galbally et al., 'Biosolid and Distillery Effluent Amendments to Irish Short
Rotation Coppiced Willow Plantations: Impacts on Groundwater Quality
and Soil', *Agricultural Water Management* (2012), http://dx.doi.org/10.1016
/j.agwat.2012.07.010.

3. R. Spinelli et al., 'Performance of a Mobile Mechanical Screen to Improve the
Commercial Quality of Wood Chips for Energy', *Bioresource Technology* 102,
no. 15 (August 2011): 7366–70, https://doi.org/10.1016/j.biortech.2011.05.002.

4. 'Compost 101', O₂Compost, https://www.o2compost.com/compost-101.aspx.

5. L.J. Sikora and M.A. Sowers, 'Effect of Temperature Control on the Composting Process', *Journal of Environmental Quality* 14, no. 3 (1985): 434–39, https://
doi.org/10.2134/jeq1985.00472425001400030025x.

6. M. Chambers et al., 'Hedgerow Harvesting Machinery Trials Report', Organic
Research Centre (2015), https://tinyurl.com/twecom.

7. Robert J. Wolton, 'The Yield and Cost of Harvesting Wood Fuel from Hedges
in the Tamar Valley and Blackdowns AONBs, South-West England', unpublished report to the Tamar Valley Area of Outstanding Natural Beauty and
Blackdown Hills AONB (2012).

Chapter 4: Woodchip for Plant Propagation

1. T. Minayeva et al., 'Peatlands and Biodiversity', in *Assessment on Peatlands,
Biodiversity and Climate Change*, F. Parish et al., eds. (Wageningen: Wetlands

International, 2008): 60–98, https://www.researchgate.net/publication /293206397_Peatlands_and_biodiversity.

2. 'Guidelines for the Specification of Quality Compost for Use in Growing Media', WRAP (February 2014), https://wrap.org.uk/resources/guide /guidelines-quality-compost-use-growing-media.

3. 'Growing Media Monitor: Trends in the Composition of UK Growing Media Supplied 2011 to 2019', Growing Media Association (2020), https://growing media.co.uk/research.html.

4. S.J. van Donk et al., 'Wood Chip Mulch Thickness Effects on Soil Water, Soil Temperature, Weed Growth and Landscape Plant Growth', *Journal of Applied Horticulture* 13, no. 2 (2011): 91–95, https://doi.org/10.37855/jah.2011.v13i02.22.

Chapter 5: Woodchip as Soil Amendment

1. Z. Liu et al., 'Effects of Biochar Application on Nitrogen Leaching, Ammonia Volatilization and Nitrogen Use Efficiency in Two Distinct Soils', *Journal of Soil Science and Plant Nutrition* 17, no. 2 (June 2017), http://doi.org/10.4067 /S0718-95162017005000037.

2. R.L. Mulvaney, S.A. Khan and T.R. Ellsworth, 'Synthetic Nitrogen Fertilizers Deplete Soil Nitrogen: A Global Dilemma for Sustainable Cereal Production', *Journal of Environmental Quality* 38, no. 6 (2009): 2295–314, http://doi.org /10.2134/jeq2008.0527.

3. E.M. Miller and T.R. Seastedt, 'Impacts of Woodchip Amendments and Soil Nutrient Availability on Understory Vegetation Establishment Following Thinning of a Ponderosa Pine Forest', *Forest Ecology and Management* 258, no. 3 (June 2009): 263–72, https://doi.org/10.1016/j.foreco.2009.04.011.

4. L. Hoagland et al., 'Orchard Floor Management Effects on Nitrogen Fertility and Soil Biological Activity in a Newly Established Organic Apple Orchard', *Biology and Fertility of Soils* 45 (2008): 11, http://doi.org/10.1007/s00374-008-0304-4.

5. R. Rynk et al., *On-Farm Composting Handbook* (Ithaca: Northeast Regional Agricultural Engineering Service, 1992), 6–13, 106–113; Sally Westaway, 'Ramial Woodchip in Agricultural Production: WOOFS Technical Guide 1', Organic Research Centre (2020), https://www.organicresearchcentre.com /wp-content/uploads/2020/12/WOOFS_TG2_Final.pdf.

6. S.A. Napieralski et al., 'Microbial Chemolithotrophy Mediates Oxidative Weathering of Granitic Bedrock', *Proceedings of the National Academy of Sciences* 116, no. 52 (26 December 2019): 26394–401, https://doi.org/10.1073/pnas.1909970117.

7. A.G. Jongmans et al., 'Rock-eating Fungi', *Nature* 389, no. 6652 (1997): 682–83, https://www.nature.com/articles/39493?foxtrotcallback=true.

8. R.D. Bardgett, P.J. Hobbs and Å. Frostegård, 'Changes in Soil Fungal:Bacterial Biomass Ratios Following Reductions in the Intensity of Management of an Upland Grassland', *Biology and Fertility of Soils* 22 (1996): 261–64, http://doi .org/10.1007/s003740050108.

9. X. Liu et al., 'Species Decline under Nitrogen Fertilization Increases Commu-
   nity-Level Competence of Fungal Diseases', *Proceedings of the Royal Society B*
   284, no. 1847 (2017), https://doi.org/10.1098/rspb.2016.2621.
10. J.G. Zaller et al., 'Glyphosate Herbicide Affects Belowground Interactions
    Between Earthworms and Symbiotic Mycorrhizal Fungi in a Model Ecosys-
    tem', *Scientific Reports* 4 (2014): 5634, https://doi.org/10.1038/srep05634.
11. 'Alternative Methods for Terminating Cover Crops', Innovative Farmers Field
    Lab, https://www.innovativefarmers.org/field-lab/?id=e05323bd-125e
    -e611-80ca-005056ad0bd4.
12. A. Gattinger et al., 'Enhanced Top Soil Carbon Stocks under Organic Farm-
    ing', *Proceedings of the National Academy of Sciences* 109, no. 44 (30 October
    2012): 18226–31, https://doi.org/10.1073/pnas.1209429109.
13. R. Eschen et al., 'Carbon Addition Alters Vegetation Composition on Ex-
    Arable Fields', *Journal of Applied Ecology* 44, no. 1 (February 2007): 95–104,
    http://doi.org/10.1111/j.1365-2664.2006.01240.x.
14. A.A. Malik et al., 'Soil Fungal:Bacterial Ratios Are Linked to Altered Carbon
    Cycling', *Frontiers in Microbiology* 7 (2016): 1247, http://doi.org/10.3389/fmicb
    .2016.01247.
15. Sally Westaway, 'Ramial Woodchip Production and Use: WOOFS Technical
    Guide 1', Organic Research Centre (November 2020), https://www.organic
    researchcentre.com/wp-content/uploads/2020/12/WOOFS_TG1_Final.pdf;
    Westaway, 'Ramial Woodchip in Agricultural Production: WOOFS Technical
    Guide 2', Organic Research Centre (November 2020), https://www.organic
    researchcentre.com/wp-content/uploads/2020/12/WOOFS_TG2_Final
    .pdf; Westaway, 'Using Ramial Woodchip as Part of a Whole Farm System:
    WOOFS Technical Guide 3', Organic Research Centre (December 2020),
    https://www.organicresearchcentre.com/wp-content/uploads/2020/12
    /WOOFS_T3_final.pdf.
16. G. Lemieux and D. Germain, 'Ramial Chipped Wood: The Clue to a Sus-
    tainable Fertile Soil', Groupe de Coordination sur les Bois Raméaux (2001),
    https://www.researchgate.net/publication/228364133_Ramial_Chipped
    _Wood_the_Clue_to_a_Sustainable_Fertile_Soil.
17. I.B. Noël, 'Le Bois Raméal Fragmenté (BRF), un nouvel élan pour l'agriculture
    bio wallonne?', *Revue Aggra* 4 (2006), http://andre.emmanuel.free.fr/brf
    /articles/aggradation4.pdf.
18. Gilles Lemieux, 'The Hidden World That Feeds Us: The Living Soil',
    Coordination Group on Ramial Wood (1996), https://cepeas.org/wp-content
    /uploads/2018/05/5-Lemieux-Living-Soil.pdf.
19. E.E. Nelson, 'Effect of Urea and Wood Shavings on Populations of Soil
    Microfungi, Especially Trichoderma Species', *Microbios* 5, no. 17 (1972): 69–72.
20. N.A. Bhat et al., 'Soil Biological Activity Contributing to Phosphorus Avail-
    ability in Vertisols under Long-Term Organic and Conventional Agricultural

Management', *Frontiers in Plant Science* 8 (2017): 1523, http://doi.org/10.3389/fpls.2017.01523.

21. T. Rouquerol, D. Bauzon, and Y. Dommergues, 'Les Ectomycorhizes et la Nutrition Azotée et Phosphatée des Arbres', Congrès DGRST (1975).

22. T. Stevanovic, 'Constituants du Bois et la Pédogenèse à Partir des BRF – Une Solution pour un Sol Durable: Mettre en Synergie Agriculture et Foresterie', *Revue Aggra* 4 (2006), http://andre.emmanuel.free.fr/brf/articles/aggradation4.pdf.

23. Charles Darwin, *The Formation of Vegetable Mould, through the Action of Worms, with Observations on Their Habits* (1881).

24. G.R. Free, 'Soil Management for Vegetable Production on Honeoye Soil with Special Reference to the Use of Hardwood Chips', *New York's Food and Life Sciences Bulletin* 2 (1971), https://ecommons.cornell.edu/handle/1813/4025.

25. Sally Westaway, 'Ramial Woodchip in Agricultural Production: WOOFS Technical Guide 2', Organic Research Centre (November 2020), https://www.organicresearchcentre.com/wp-content/uploads/2020/12/WOOFS_TG2_Final.pdf.

26. F. Alliaume et al., 'Modelling Soil Tillage and Mulching Effects on Soil Water Dynamics in Raised-bed Vegetable Rotations', *European Journal of Agronomy* 82, Part B (January 2017): 268–81, http://doi.org/10.1016/j.eja.2016.08.011.

27. Linda Chalker-Scott, 'Hugelkultur: What Is It, and Should It Be Used on Home Gardens?', *Washington State University Extension* (August 2017), http://hdl.handle.net/2376/12233.

28. 'Innovative Uses of Compost: Bioremediation and Pollution Prevention', United States Environmental Protection Agency (October 1997), https://www.epa.gov/sites/production/files/2015-08/documents/bioremed.pdf.

29. Ugochukwu C. Okafor and Amechi S. Nwankwegu, 'Effect of Woodchips on Bioremediation of Crude Oil-Polluted Soil', *British Microbiology Research Journal* 15, no. 4 (2016), http://doi.org/10.9734/BMRJ/2016/27027.

30. K. Robichaud Girard et al., 'Local Fungi, Willow and Municipal Compost Effectively Remediate Petroleum-contaminated Soil in the Canadian North," *Chemosphere* 220 (April 2019): 47–55, https://doi.org/10.1016/j.chemosphere.2018.12.108.

31. W. Spisak et al., 'Using Wood Chips for the Protection of Plants and Soil from the Harmful Effects of Road Salt', *European Journal of Wood and Wood Products* 78 (2020): 1209–19, https://link.springer.com/article/10.1007/s00107-020-01563-4; J. Miller et al., 'Surface Soil Salinity and Soluble Salts after 15 Applications of Composted or Stockpiled Manure with Straw or Wood-Chips', *Compost Science & Utilization* 25, no. 1 (2017): 36–47, http://doi.org/10.1080/1065657X.2016.1176968.

32. M. Shaygan et al., 'The Effect of Soil Physical Amendments on Reclamation of a Saline-sodic Soil: Simulation of Salt Leaching Using HYDRUS-1D', *Soil Research* 56, no. 8 (2018): 829–45, https://doi.org/10.1071/SR18047.

33. S. Brown et al., 'Effect of Biosolids Processing on Lead Bioavailability in an Urban Soil', *Journal of Environmental Quality* 32, no. 1 (January 2003): 100–8, https://doi.org/10.2134/jeq2003.1000.

34. L.A. Schipper et al., 'Denitrifying Bioreactors: An Approach for Reducing Nitrate Loads to Receiving Waters', *Ecological Engineering* 36, no. 11 (November 2010): 1532–43, http://doi.org/10.1016/j.ecoleng.2010.04.008.

35. E.C. Wolf, E. Rejmánková and D.J. Cooper, 'Wood Chip Soil Amendments in Restored Wetlands Affect Plant Growth by Reducing Compaction and Increasing Dissolved Phenolics', *Restoration Ecology* 27, no. 5 (November 2019): 1128–36, https://doi.org/10.1111/rec.12942.

36. B.C. Scharenbroch and G.W. Watson, 'Wood Chips and Compost Improve Soil Quality and Increase Growth of *Acer rubrum* and *Betula nigra* in Compacted Urban Soil', *Arboriculture and Urban Forestry* 40, no. 6 (2014): 319–31, https://www.researchgate.net/publication/288097189_Wood_chips_and_compost_improve_soil_quality_and_increase_growth_of_Acer_rubrum_and_Betula_nigra_in_compacted_urban_soil.

37. J. León et al., 'Effectiveness of Wood Chips Cover at Reducing Erosion in Two Contrasted Burnt Soils', *Zeitschrift für Geomorphologie: Supplementary Issue* 57, no. 1 (2013): 27–37, http://doi.org/10.1127/0372-8854/2012/S-00086.

Chapter 6: Mulches

1. D.A. Shaw, D.R. Pittenger and M. McMaster, 'Water Retention and Evaporative Properties of Landscape Mulches', *Irrigation Association* (2005), https://ucanr.edu/sites/UrbanHort/files/80238.pdf.

2. S.J. van Donk et al., 'Wood Chip Mulch Thickness Effects on Soil Water, Soil Temperature, Weed Growth and Landscape Plant Growth', *Journal of Applied Horticulture* 13, no. 2 (2011): 91–5, http://doi.org/10.37855/jah.2011.v13i02.22.

3. Nazim Gruda, 'The Effect of Wood Fiber Mulch on Water Retention, Soil Temperature and Growth of Vegetable Plants', *Journal of Sustainable Agriculture* 32, no. 4 (2008): 629–43, http://doi.org/10.1080/10440040802395049.

4. B.C. Scharenbroch and G.W. Watson, 'Wood Chips and Compost Improve Soil Quality and Increase Growth of *Acer rubrum* and *Betula nigra* in Compacted Urban Soil', *Arboriculture and Urban Forestry* 40, no. 6 (2014): 319–31, https://www.researchgate.net/publication/288097189_Wood_chips_and_compost_improve_soil_quality_and_increase_growth_of_Acer_rubrum_and_Betula_nigra_in_compacted_urban_soil.

5. David Granatstein and Kent Mullinix, 'Mulching Options for Northwest Organic and Conventional Orchards', *HortScience* 43, no. 1 (2008): 45–50, http://doi.org/10.21273/HORTSCI.43.1.45.

6. K.D. Patten et al., 'Growth and Yield of Rabbiteye Blueberry as Affected by Orchard Floor Management Practices and Irrigation Geometry', *Journal of the*

American Society for Horticultural Science (1989), https://agris.fao.org
/agris-search/search.do?recordID=US19900024718.

7. Scharenbroch and Watson, 'Wood Chips and Compost Improve Soil Quality', 319–31.

8. van Donk et al., 'Wood Chip Mulch Thickness Effects', 91–5.

9. van Donk et al., 'Wood Chip Mulch Thickness Effects', 91–5.

10. K.M. Greenly and D.A. Rakow, 'The Effect of Wood Mulch Type and Depth on Weed and Tree Growth and Certain Soil Parameters', *Journal of Arboriculture* 21, no. 5 (1995): 225–32, https://www.semanticscholar.org/paper /THE-EFFECT-OF-WOOD-MULCH-TYPE-AND-DEPTH-ON-WEED -AND-Greenly-Rakow/d4251ffaea999bd195e4a396e401546e1f42f4d0.

11. Pavol Findura et al., 'Evaluation of the Effects of Allelopathic Aqueous Plant Extracts, as Potential Preparations for Seed Dressing, on the Modulation of Cauliflower Seed Germination', *Agriculture* 10, no. 4 (2020), http://doi .org/10.3390/agriculture10040122.

12. B. Rathinasabapathi, J. Ferguson and M. Gal, 'Evaluation of Allelopathic Potential of Wood Chips for Weed Suppression in Horticultural Production Systems', *HortScience* 40, no. 3 (June 2005), https://www.researchgate.net /publication/267833486_Evaluation_of_Allelopathic_Potential_of_Wood _Chips_for_Weed_Suppression_in_Horticultural_Production_Systems.

13. Glynn Percival, 'What's New in Plant Protection', Bartlett Tree Research Laboratory, https://www.ltoa.org.uk/docs/LTOA-Mulches-Glynn_Percival.pdf.

14. M.W. Brown and T. Tworkoski, 'Pest Management Benefits of Compost Mulch in Apple Orchards', *Agriculture, Ecosystems & Environment* 103, no. 3 (August 2004): 465–72, http://doi.org/10.1016/j.agee.2003.11.006.

15. J. Downer, B. Faber and J. Menge, 'Factors Affecting Root Rot Control in Mulched Avocado Orchards', *HortTechnology* 12, no. 4 (2002): 601–5, http:// doi.org/10.21273/HORTTECH.12.4.601.

16. R.D. Harter and G. Stotzky, 'Formation of Clay Protein Complexes', *Soil Science Society of America Journal* 35, no. 3 (1971): 383–89, https://doi.org /10.2136/sssaj1971.03615995003500030019x.

17. E.B. Himelick and G.W. Watson, 'Reduction of Oak Chlorosis with Wood Chip Mulch Treatments', *Journal of Arboriculture* 16, no. 10 (1990): 275–78, http://joa .isa-arbor.com/request.asp?JournalID=1&ArticleID=2385&Type=2.

18. G.C. Percival, 'Influence of Pure Mulches on Suppressing *Phytophthora* Root Rot Pathogens', *Journal of Environmental Horticulture* 31, no. 4 (2013): 221–26, https://doi.org/10.24266/0738-2898.31.4.221.

19. James Downer et al., 'The Effect of Yard Trimmings as a Mulch on Growth of Avocado and Avocado Root Rot Caused by *Phytophthora cinnamomi*', *California Avocado Society 1999 Yearbook* 83 (1999): 87–104, https://www.researchgate .net/publication/237540772_The_Effect_of_Yard_Trimmings_as_a_Mulch _on_Growth_of_Avocado_and_Avocado_Root_Rot_Caused_by_Phytophthora _cinnamomi.

20. T. Senaratna et al., 'Acetyl Salicylic Acid (Aspirin) and Salicylic Acid Induce Multiple Stress Tolerance in Bean and Tomato Plants', *Plant Growth Regulation* 30 (2000): 157–61, https://doi.org/10.1023/A:1006386800974.

21. L.A. Lacey and T. Unruh, 'Biological Control of Codling Moth (Cydia pomonella, Lepidoptera: Tortricidae) and Its Role in Integrated Pest Management, with Emphasis on Entomopathogens', *Vedalia* 12, no. 1 (2005): 33–60, https://www.researchgate.net/publication/286336481_Biological_control _of_codling_moth_Cydia_pomonella_Lepidoptera_Tortricidae_and_its _role_in_integrated_pest_management_with_emphasis_on_entomopathogens.

22. D.C. Lloyd, 'Memorandum on Natural Enemies of the Codling Moth, Carpocapsa pomonella (L.)', CIBC Report (1960).

23. L.A. Lacey et al., 'Use of Entomopathogenic Nematodes (Steinernematidae) in Conjunction with Mulches for Control of Overwintering Codling Moth (Lepidoptera: Tortricidae)', *Journal of Entomological Science* 41, no. 2 (2006): 107–19, http://doi.org/10.18474/0749-8004-41.2.107.

24. J. Pietikäinen, M. Pettersson, and E. Bååth, 'Comparison of Temperature Effects on Soil Respiration and Bacterial and Fungal Growth Rates', *FEMS Microbiology Ecology* 52, no. 1 (March 2005): 49–58, https://doi.org/10.1016 /j.femsec.2004.10.002.

25. van Donk et al., 'Wood Chip Mulch Thickness Effects', 91–5; J. Cox, 'Comparison of Plastic Weedmat and Woodchip Mulch on Low Chill Blueberry Soil in New South Wales, Australia', *Acta Horticulturae* 810, 475–82, http://doi .org/10.17660/ActaHortic.2009.810.62.

26. K.M. Greenly and D.A. Rakow, 'The Effect of Wood Mulch Type and Depth on Weed and Tree Growth and Certain Soil Parameters', *Journal of Arboriculture* 21, no. 5 (1995): 225–32, https://www.semanticscholar.org/paper/THE-EFFECT -OF-WOOD-MULCH-TYPE-AND-DEPTH-ON-WEED-AND-Greenly-Rakow /d4251ffaea999bd195e4a396e401546e1f42f4d0.

27. G. Amoroso et al., 'Effect of Mulching on Plant and Weed Growth, Substrate Water Content, and Temperature in Container-grown Giant Arborvitae', *Hort-Technology* 20, no. 6 (2010): 957–62, http://doi.org/10.21273/HORTSCI.20.6.957.

28. J.A. Cox, S. Morris and T. Dalby, 'Woodchip or Weedmat? A Comparative Study on the Effects of Mulch on Soil Properties and Blueberry Yield', *Acta Horticulturae* 1018 (2014): 369–74, http://doi.org/10.17660/ActaHortic.2014.1018.39.

29. I.A. Merwin et al., 'Comparing Mulches, Herbicides, and Cultivation as Orchard Groundcover Management Systems', *HortTechnology* 5, no. 2 (1995): 151–8, http://doi.org/10.21273/HORTTECH.5.2.151.

30. L. Chalker-Scott, A. Cahill and K. Ewing, 'Wood-chip Mulch Improves Plant Survival and Establishment at No-Maintenance Restoration Site', *Ecological Restoration* 23 (2005): 212–13, https://www.researchgate.net/publication /303445066_Wood-chip_mulch_improves_plant_survival_and_establishment _at_no-maintenance_restoration_site.

Chapter 7: Mushrooms in Woodchip

1. M.N. Hasan et al., 'Performance of Oyster Mushroom on Different Pretreated Substrates', *International Journal of Sustainable Crop Production* 5, no. 4 (2010): 16–24, https://www.researchgate.net/publication/259192028_Performance _of_Oyster_mushroom_on_different_pretreated_substrate.

2. J. Carrasco et al., 'Supplementation in Mushroom Crops and Its Impact on Yield and Quality', *AMB Express* 8, no. 1 (2018): 146, http://doi.org/10.1186 /s13568-018-0678-0.

3. T. Roenneberg and M. Merrow, 'Seasonality and Photoperiodism in Fungi', *Journal of Biological Rhythms* 16, no. 4 (2001): 403–14, http://doi.org/10.1177 /074873001129001999.

4. S.C. Croan, 'Bioconversion of Conifer Wood Chips into Specialty Mushroom Producing Fungal Growth', *Mushrooms International* 90 (2002): 7–11, https:// www.fpl.fs.fed.us/documnts/pdf2002/croan02a.pdf; S.C. Croan, 'Utilization of Treated Conifer Wood Chips by Pleurotus (Fr.) P. Karst. Species for Cultivating Mushrooms', *Mushrooms International* 91 (2003): 4–7, http://www .fungifun.org/docs/mushrooms/Utilization%20of%20treated%20conifer%20 wood%20chips%20by%20pleurotus%20species%20for%20cultivating%20 mushrooms.pdf.

Chapter 8: Building Carbon with Woodchip

1. R. Lal, 'Global Potential of Soil Carbon Sequestration to Mitigate the Green-house Effect', *Critical Reviews in Plant Sciences* 22, no. 2 (2003): 151–84, https:// www.tandfonline.com/doi/abs/10.1080/713610854; J. Sanderman et al., 'Soil Carbon Debt of 12,000 Years of Human Land Use', *Proceedings of the National Academy of Sciences* 114, no. 36 (2017): 9575–80, http://doi.org/10.1073 /pnas.1706103114.

2. I.M. Lubbers et al., 'Greenhouse-gas Emissions from Soils Increased by Earthworms', *Nature Climate Change* 3 (2013): 187–94, https://doi.org/10.1038 /nclimate1692.

3. T.A. Ontl and L.A. Schulte, 'Soil Carbon Storage', *Nature Education Knowledge* 3, no. 10 (2012), https://www.researchgate.net/publication/313189912_Soil _carbon_storage.

4. R.J. Zomer et al., 'Global Sequestration Potential of Increased Organic Carbon in Cropland Soils', *Scientific Reports* 7 (2017): 15554, https://doi.org/10.1038 /s41598-017-15794-8.

5. K.E. Clemmensen et al., 'Roots and Associated Fungi Drive Long-Term Carbon Sequestration in Boreal Forest', *Science* 339, no. 6127 (2013): 1615–18, http:// doi.org/10.1126/science.1231923.

6. C.E. Prescott et al., 'Surplus Carbon Drives Allocation and Plant–Soil Interac-tions', *Trends in Ecology & Evolution* 35, no. 12 (2020): 1110–18, https://www .cell.com/trends/ecology-evolution/fulltext/S0169-5347(20)30222-6.

7. Kathleen K. Treseder, 'A Meta-analysis of Mycorrhizal Responses to Nitrogen, Phosphorus, and Atmospheric CO_2 in Field Studies', *New Phytologist* 164, no. 2 (2004): 347–55, https://escholarship.org/content/qt8783k14r/qt8783k14r.pdf.

8. J. Six et al., 'Bacterial and Fungal Contributions to Carbon Sequestration in Agroecosystems', *Soil Science Society of America Journal* 70, no. 2 (2006): 555–69, http://doi.org/10.2136/sssaj2004.0347.

9. A.A. Malik et al., 'Soil Fungal:Bacterial Ratios Are Linked to Altered Carbon Cycling', *Frontiers in Microbiology* 7 (2016): 1247, http://doi.org/10.3389/fmicb.2016.01247.

10. Marta Prada et al., 'Carbon Sequestration for Different Management Alternatives in Sweet Chestnut Coppice in Northern Spain', *Journal of Cleaner Production* 135, no. 1 (2016): 1161–69, https://doi.org/10.1016/j.jclepro.2016.07.041; Nathália Faria da Silva et al., 'Yield and Nutrient Demand and Efficiency of Eucalyptus under Coppicing Regime', *Forests* 11, no. 8 (2020): 852, http://doi.org/10.3390/f11080852.

11. Christopher Poeplau and Don Axel, 'Carbon Sequestration in Agricultural Soils via Cultivation of Cover Crops – A Meta-Analysis,' *Agriculture, Ecosystems & Environment* 200 (2015): 33–41, https://www.sciencedirect.com/science/article/abs/pii/S0167880914004873?via%3Dihub.

12. B.M. Shrestha et al., 'Effects of Crop Rotation, Crop Type and Tillage on Soil Organic Carbon in a Semiarid Climate', *Canadian Journal of Soil Science* 93, no. 1 (2013): 137–46, https://cdnsciencepub.com/doi/10.4141/cjss2012-078.

13. Emre Gençer et al., 'Sustainable Production of Ammonia Fertilizers from Biomass', *Biofuels Bioproducts & Biorefining* 14, no. 4 (2020), https://onlinelibrary.wiley.com/doi/full/10.1002/bbb.2101.

14. J.W. White, 'Soil Organic Matter and Manurial Treatments', *Agronomy Journal* 19, no. 5 (1927): 389–96, http://doi.org/10.2134/agronj1927.00021962001900050004x; R.L. Mulvaney et al., 'Synthetic Nitrogen Fertilizers Deplete Soil Nitrogen: A Global Dilemma for Sustainable Cereal Production', *Journal of Environmental Quality* 38, no. 6 (2009): 2295–314, https://acsess.onlinelibrary.wiley.com/doi/10.2134/jeq2008.0527.

15. Ahmed Sheikh et al., 'Ammonium Nitrate-impregnated Woodchips: A Slow-release Nitrogen Fertilizer for Plants', *Journal of Wood Science* 57 (2011): 295–301, https://jwoodscience.springeropen.com/articles/10.1007/s10086-011-1178-x.

16. A.J. Kuthubutheen and G.J.F. Pugh, 'The Effects of Fungicides on Soil Fungal Populations', *Soil Biology and Biochemistry* 11, no. 3 (1979): 297–303, https://www.sciencedirect.com/science/article/abs/pii/0038071779900750; M. Wainwright and G.J.F. Pugh, 'Effect of Fungicides on the Numbers of Microorganisms and Frequency of Cellulolytic Fungi in Soils', *Plant and Soil* 43, no. 3 (1975): 561–72, https://link.springer.com/article/10.1007/BF01928519.

17. Valerie Wilkinson and R.L. Lucas, 'Effects of Herbicides on the Growth of Soil Fungi', *New Phytologist* 68, no. 3 (1969): 709–19, https://nph.onlinelibrary .wiley.com/doi/abs/10.1111/j.1469-8137.1969.tb06475.x.

18. A.Y. Abdel-Mallek, M.I. Abdel-Kader and A.M. Shonkeir, 'Effect of Glyphosate on Fungal Population, Respiration and the Decay of Some Organic Matters in Egyptian Soil', *Microbiological Research* 149, no. 1:69–73, http://doi.org /10.1016/S0944-5013(11)80139-4.

19. XiaoLi Bu et al., 'Effect of Biochar on Seed Germination and Seedling Growth of *Robinia pseudoacacia L.* in Karst Calcareous Soils', *Communications in Soil Science and Plant Analysis* 51, no. 3 (2020): 352–63, http://doi.org/10.1080/00103 624.2019.1709484; Omer Suha Uslu et al., 'Walnut Shell Biochar Increases Seed Germination and Early Growth of Seedlings of Fodder Crops', *Agriculture* 10, no. 10 (2020): 427, https://www.mdpi.com/2077-0472/10/10/427/htm.

20. O.Y. Yu et al., 'Impact of Biochar on the Water Holding Capacity of Loamy Sand Soil', *International Journal of Energy Environmental Engineering* 4 (2013), https://link.springer.com/article/10.1186/2251-6832-4-44.

21. David A. Wardle, Marie-Charlotte Nilsson and Olle Zackrisson, 'Fire-Derived Charcoal Causes Loss of Forest Humus', *Science* 320, no. 5876: 629, http:// doi.org/10.1126/science.1154960.

22. Andressa M. Freitas, Vimala D. Nair and Willie G. Harris, 'Biochar as Influenced by Feedstock Variability: Implications and Opportunities for Phosphorus Management', *Frontiers in Sustainable Food Systems* 4 (2020), https://www.frontiersin.org/articles/10.3389/fsufs.2020.510982/full; Vimala D. Nairet et al., 'Biochar in the Agroecosystem–Climate-Change–Sustainability Nexus', *Frontiers in Plant Science* 8 (2017), https://www.frontiersin.org/articles /10.3389/fpls.2017.02051/full.

23. Hans-Peter Schmidt et al., 'The Use of Biochar in Animal Feeding', *PeerJ* (2019): 7373, http://doi.org/10.7717/peerj.7373 https://peerj.com/articles/7373/.

24. J.A. Micales and K.E. Skog, 'The Decomposition of Forest Products in Landfills', *International Biodeterioration & Biodegradation* 39, nos. 2–3 (1997): 145–58, https://doi.org/10.1016/S0964-8305(97)83389-6.

25. N. Zeng, 'Carbon Sequestration via Wood Burial', *Carbon Balance and Management* 3 (2008): 1, https://cbmjournal.biomedcentral.com/articles /10.1186/1750-0680-3-1.

26. Carly Whittaker et al., 'Dry Matter Losses and Methane Emissions during Wood Chip Storage: The Impact on Full Life Cycle Greenhouse Gas Savings of Short Rotation Coppice Willow for Heat', *BioEnergy Research* 9 (2016), http://doi.org/10.1007/s12155-016-9728-0.

27. Marcella Guarino et al., 'Evaluation of Simplified Covering Systems to Reduce Gaseous Emissions from Livestock Manure Storage', *Transactions of the ASABE (American Society of Agricultural and Biological Engineers)* 49, no. 3: 737, http://doi.org/10.13031/2013.20481.

INDEX

Note: Italicised page numbers indicate a photo or caption. A lowercase *i* indicates the photo insert.

ABOUT THE AUTHOR

Ruth Raskin

Ben Raskin has worked in horticulture for more than twenty-five years and has a wide range of experience in both practical commercial growing and wider policy and advocacy work.

As head of horticulture and agroforestry for the Soil Association, he provides growers at all levels of production with technical, marketing, policy, supply chain and networking support. He is currently implementing a 200-acre silvopastoral agroforestry planting in Wiltshire, England.

Ben is the author of several previous books on gardening, including *Zero-Waste Gardening* (2021) and *The Community Gardening Handbook* (2017), as well as three volumes of the Grow Together Guides, aimed at families with young children: *Compost*, *Grow* and *Bees, Bugs and Butterflies*.

Additionally, Ben co-chairs the Defra Edibles Horticulture Roundtable, and sits on the boards of the Organic Growers Alliance and Community Supported Agriculture Network UK.

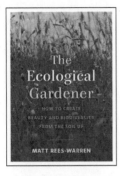
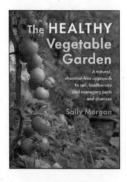
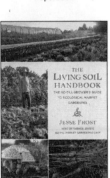
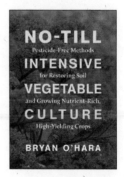

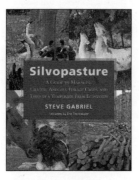
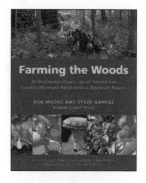
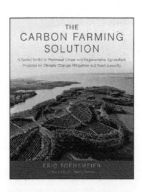
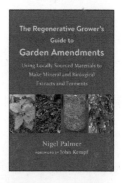